MEDICAL RADIOLOGY
Diagnostic Imaging

David Wilson (Ed.)

Paediatric Musculoskeletal Disease

With an Emphasis on Ultrasound

With Contributions by

G. Allan · S. Bianchi · R. Cheung · C. Groves · C. Martinoli · P. J. O'Connor
P. Thomà · M. Valle · D. Wilson

Foreword by
A. L. Baert

With 88 Figures in 141 Separate Illustrations, 11 in Color and 3 Tables

Springer

David Wilson, FRCP, FRCR
Department of Radiology
Nuffield Orthopaedic Centre
NHS Trust
Windmill Road, Headington
Oxford OX3 7LD
UK

Medical Radiology · Diagnostic Imaging and Radiation Oncology
Series Editors: A. L. Baert · L. W. Brady · H.-P. Heilmann · M. Molls · K. Sartor

Continuation of Handbuch der medizinischen Radiologie
Encyclopedia of Medical Radiology

Library of Congress Control Number: 2003060242

ISBN 3-540-66828-4 Springer Berlin Heidelberg New York

Springer is part of Springer Science+Business Media

http//www.springeronline.com

Printed in Germany

Medical Editor: Dr. Ute Heilmann, Heidelberg
Desk Editor: Ursula N. Davis, Heidelberg
Production Editor: Kurt Teichmann, Mauer
Cover-Design and Typesetting: Verlagsservice Teichmann, Mauer

Printed on acid-free paper – 21/3150xq – 5 4 3 2 1 0

Dedication

To all your children
and
to the young people in our family:
Emma, Phil, Iain, Neil,
Rachael, Siân, Natalie and *Jack.*

Foreword

During recent years the important role of ultrasound in the diagnosis and management of musculoskeletal diseases has become well established. The specific strengths of this modality become quite evident in the examination of children because of the absence of ionising radiation and the close interaction between examiner and patient.

This volume highlights the role of ultrasound as a primary or adjuvant diagnostic modality in a range of musculoskeletal disorders and describes its contribution for problem solving and for better patient management.

The eminently readable text is completed by superb illustrations.

The editor, D. Wilson, is a world-renowned expert in the field with a great dedication and interest in paediatric ultrasound. The authors of individual chapters were invited to participate because of their long-standing experience and major contributions to the radiological literature on the topic.

I would like to thank the editor and the authors and congratulate them most sincerely for their superb efforts that have resulted in this outstanding volume.

This book will be of great value not only for general and paediatric radiologists but also for paediatricians and paediatric orthopaedic surgeons. It will provide all of these specialists with state-of-the-art information on a narrow but fascinating medical field.

I am confident that it will meet the same success with readers as previous volumes in this series.

Leuven ALBERT L. BAERT

Preface

When I started examining patients with ultrasound for musculoskeletal disorders we were still using static "B" scanners. CT was a new invention and MRI did not exist. Whilst my contemporaries were enthusiastically specialising in the use of nuclear medicine and ultrasound, I chose to take an interest and eventually a full-time specialisation in a system rather than a machine. The principal strength of this choice is that I use all imaging methods and hopefully have insight into their advantages and weaknesses in each potential application. It does mean that I have little knowledge of what goes on in the abdomen, chest or brain and that, for better or worse, I will have to work with orthopaedic surgeons and rheumatologists for the remainder of my career.

Reading textbooks on imaging I am struck by the sometimes tortuous reasoning that others employ to explain why their preferred technique is useful to make a diagnosis in all illnesses. They are at times blinkered to the strengths of a rival method and often overlook the potential for correlation of the signs on more than one technique. My view is that musculoskeletal radiologists should be skilled in the interpretation of plain films, ultrasound, nuclear medicine, CT and MRI. To miss out one of these is to be like a gardener with a fork and rake but no spade. There is no place for rivalry as all are needed to carry out an accurate, safe and effective practice.

I was asked to edit a text on paediatric musculoskeletal imaging. I have chosen to emphasise the role of ultrasound, which is very well suited to the examination of children, who commonly will not lie still, require reassurance by personal contact and need a non-invasive and comfortable method of examination. We are all aware of the dangers of ionising radiation and the potential for fear and anxiety when inside the bore of an MR system. However, CT and MRI are powerful means of making a diagnosis, and sedation or anaesthesia are practical options.

Clinicians, therapists and diagnosticians need to understand the strengths and weaknesses of each imaging method with knowledge of the alternative methods.

This book describes a range of practical problems in diagnosis and therapy where ultrasound has a significant role to play. In some circumstances ultrasound is the most effective means of diagnosis, as in the assessment of developmental dysplasia or stress-testing the integrity of a tendon. We consider the support ultrasound lends to other methods, perhaps confirming a diagnosis, for example the detection of inflammatory synovitis in juvenile arthropathy. Lastly, we discuss where ultrasound is a weaker alternative to other more comprehensive methods, such as in the detection of periosteal reaction over an area of osteomyelitis.

We do not cover those areas of musculoskeletal practice where ultrasound has no current role. For example, CT and MRI are the most effective means of making a diagnosis of hindfoot coalition, but ultrasound cannot show us the deep structures.

Ultrasound machines vary widely in their ability to realise images of superficial lesions. In musculoskeletal applications most lesions are near the surface and care must be taken to select the best equipment . Even the latest machines may fail in this area, and thorough knowledge of the technology is important to guide our patients to effective diagnosis and treatment.

Ultrasound, CT and MRI demand a wide knowledge of anatomy, including variations of normal. Ultrasound is a dynamic examination, and the examiner must also understand the biomechanics and function of the regions studied.

The authors of the chapters in this book are all renowned musculoskeletal imaging specialists. They have been briefed to discuss all the imaging appropriate to a suspected diagnosis, and we hope that the reader will gain an understanding of where each method fits into a modern practice.

Oxford DAVID WILSON

Contents

1 **Congenital and Developmental Disorders**
David Wilson and Ruth Cheung 1

1.1 Introduction 1

1.2 Developmental Dysplasia of the Hip 1
1.2.1 Clinical Background 1
1.2.2 Role of Imaging in Detection 2
1.2.3 Role of Imaging in Treatment 4
1.2.4 Potential Developments 5

1.3 Focal Defects 7
1.3.1 Clinical Background 7
1.3.2 Role of Imaging 8
1.3.3 Potential Developments 9

1.4 Talipes Equinovarus 9
1.4.1 Clinical Background 9
1.4.2 Role of Imaging 9
1.4.3 Potential Developments 10

1.5 Neural Tube Defects 10
1.5.1 Clinical Background 10
1.5.2 Role of Imaging 11
1.5.3 Potential Developments 16

References and Further Reading 16

2 **Trauma and Sports-related Injuries**
Philip J. O'Connor and Clare Groves 19

2.1 General Principles 19
2.1.1 Biomechanics 19
2.1.2 Imaging 20

2.2 Acute Trauma 20
2.2.1 Acute Fracture Patterns 20
2.2.2 Foreign Bodies 24
2.2.3 Haematoma 25
2.2.4 Muscles and Tendons 26

2.3 Chronic Trauma 28
2.3.1 Chronic Fracture Patterns 28

2.4 The Osteochondroses 32
2.4.1 Osteochondritis Dissecans 32
2.4.2 Panner's Disease 33
2.4.3 Medial Epicondylitis (Little League Elbow) 34

2.5 Accessory Ossicles . . . 34
2.6 Bursae . . . 34
2.7 Summary . . . 37
References and Further Reading . . . 37

3 Ultrasonography of Tendons and Ligaments
Maura Valle, Stefano Bianchi, Paolo Tomà, and Carlos Martinoli . . . 39

3.1 Introduction . . . 39
3.2 Normal Anatomy . . . 39
3.2.1 Tendons . . . 39
3.2.2 Ligaments . . . 40
3.3 Examination Techniques and Normal Imaging Findings . . . 40
3.3.1 US . . . 40
3.3.2 MR imaging . . . 42
3.4 Tendon Abnormalities . . . 43
3.4.1 Overuse Injuries . . . 43
3.4.2 Avulsion Injuries . . . 43
3.4.3 Snapping Hip . . . 46
3.4.4 Degenerative and Inflammatory Conditions . . . 47
3.5 Ligament Abnormalities . . . 50
3.6 Conclusion . . . 51
References and Further Reading . . . 51

4 Inflammatory Disorders
David Wilson . . . 53

4.1 Introduction . . . 53
4.2 Irritable Hip . . . 53
4.2.1 Clinical Background . . . 53
4.2.2 Role of Imaging in Detection . . . 54
4.2.3 Role of Imaging in Treatment . . . 57
4.2.4 Management Protocol . . . 57
4.2.5 Potential Developments . . . 58
4.3 Synovitis . . . 58
4.3.1 Clinical Background . . . 58
4.3.2 Role of Imaging . . . 58
4.3.3 Potential Developments . . . 59
4.4 Osteomyelitis and Soft-tissue Infection . . . 59
4.4.1 Clinical Background . . . 59
4.4.2 Role of Imaging in Detection . . . 60
4.4.3 Role of Imaging in Treatment . . . 64
4.4.4 Potential Developments . . . 64
References and Further Reading . . . 64

5 Soft Tissue Tumours in Children
Gina Allen 67

5.1 Introduction. 67
5.2 US 67
5.2.1 Plain Radiographs. 69
5.2.2 Magnetic Resonance Imaging 69
5.3 Computed Tomography 70
5.4 Nuclear Medicine 70
5.5 Disease characteristics 71
5.5.1 Benign Lesions Seen On US. 71
5.5.2 Vascular Anomalies. 76
5.6 Potential Developments 82
References and Further Reading. 82

6 Interventional Techniques
David Wilson. 85

6.1 Introduction. 85
6.2 Biopsy 85
6.2.1 Soft Tissue Masses. 85
6.2.2 Bone Masses. 87
6.3 Aspiration. 88
6.4 Local Anaesthetic Blocks 89
6.5 Osteoid Osteoma Ablation. 89
References and Further Reading. 90

Subject Index 93

List of Contributors. 98

1 Congenital and Developmental Disorders

David Wilson and Ruth Cheung

CONTENTS

1.1 Introduction *1*
1.2 Developmental Dysplasia of the Hip *1*
1.2.1 Clinical Background *1*
1.2.2 Role of Imaging in Detection *2*
1.2.2.1 US Methods *3*
1.2.3 Role of Imaging in Treatment *4*
1.2.4 Potential Developments *5*
1.3 Focal Defects *7*
1.3.1 Clinical Background *7*
1.3.2 Role of Imaging *8*
1.3.3 Potential Developments *9*
1.4 Talipes Equinovarus *9*
1.4.1 Clinical Background *9*
1.4.2 Role of Imaging *9*
1.4.3 Potential Developments *10*
1.5 Neural Tube Defects *10*
1.5.1 Clinical Background *10*
1.5.2 Role of Imaging *11*
1.5.3 Potential Developments *16*
References and Further Reading *16*

1.1 Introduction

There are a large number of congenital birth defects that affect the spine and appendicular skeleton. They range from isolated defects affecting one part of the body to complex syndromes with several body systems involved. In practice most patients are atypical and some may cynically suggest that each case is a new syndrome. However, there are real reasons for giving as accurate a description as possible. Prognosis and outcome may be predictable and there is likely to be concern about the type of inheritance. Geneticists will look for as precise a diagnosis as possible and radiology, especially plain films, is part of that process (Fig. 1.1).

The dysplasias that predominantly involve the skeleton may be classified in a variety of ways, but the commonest is to define the part of bone most affected, epiphysis, metaphysis or diaphysis. Subgroups include the region of the skeleton most affected or other nonskeletal disorders. For example, spondyloepiphyseal dysplasia is a condition that affects the epiphyses and the spine.

There are a good number of texts that comprehensively describe syndromes that affect the musculoskeletal system and the reader is referred to them for the analysis of a particular case. In this chapter, we deal with those disorders where the imaging has a particular pivotal role in management and where ultrasound (US) has a special value.

1.2 Developmental Dysplasia of the Hip

1.2.1 Clinical Background

Developmental dysplasia of the hip (DDH) is a diagnosis made when the infant's hip is either abnormally shallow or even dislocated at birth but also when a shallow hip fails to mature to one that is mechanically stable. Its cause is not fully understood. Although there is a genetic predisposition, there is also evidence that abnormal stress on the hip in the later stages of pregnancy may lead to modelling deformity [1]. If untreated, a full dislocation will lead to the child failing to walk normally at around one year of age. A shallow and potentially unstable hip may not cause any symptoms until much later in life when the abnormal stresses lead to an acetabular labral tear or premature osteoarthritis. DDH diagnosed in infancy, by clinical examination and plain film analysis, is reported to occur between one and three times per thousand live births; the incidence of shallow or dysplastic acetabulae is much more frequent

D. Wilson, FRCP, FRCR
R. Cheung, FRCR
Department of Radiology, Nuffield Orthopaedic Centre, NHS Trust, Windmill Road, Headington, Oxford, OX3 7LD, UK

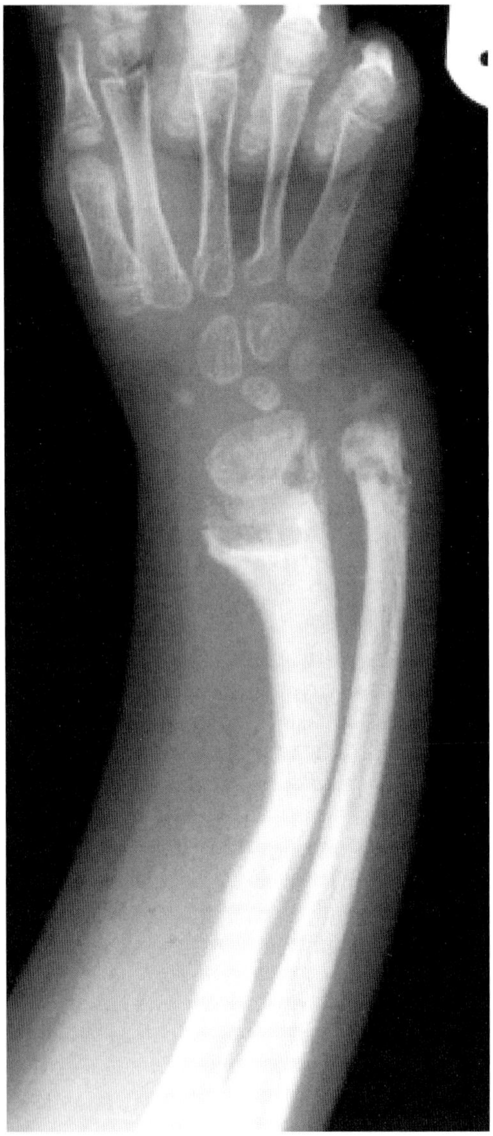

Fig. 1.1 Plain film of the forearm of a child with metaphyseal chondrodysplasia. This examination is part of a full skeletal survey.

[2]. It is difficult to identify statistics to support this comment, but experience suggests that persisting shallow acetabulae are at least ten times more common. Whilst many of these children will remodel and spontaneously recover stability, some will fail to mature properly and require a variety of complex surgical procedures [3]. It has been argued that around one-tenth of hip replacements are performed for premature osteoarthritis secondary to mild or subclinical hip dysplasia.

The goals of diagnosis and treatment are to permit affected children to walk normally and to prevent premature degeneration. We consider detection and treatment separately.

1.2.2 Role of Imaging in Detection

Most developed countries have established clinical screening methods to detect children with dislocated or dislocatable hips and there are advocates of this as the sole screening test [4]. The manoeuvres of Ortolani and Barlow are effective in detecting around 74% of cases of dislocation or subluxation that may be demonstrated on imaging. The level of training and experience required to accurately perform these tests is substantial, and sadly the task is often placed in the hands of the more junior members of the team. There are undoubtedly occasions when a child with DDH is overlooked when a clinical abnormality might have been detected by a more experienced clinician. Training and audit of practice are crucial, but even in the best hands there will be errors, as clinical manoeuvres alone are not capable of detecting every case. Indeed it is also likely that some stable hips become unstable, and if the timing of the clinical examination does not coincide with this developing problem then a child may miss the chance of early treatment that could potentially limit or reverse the process.

The need for early diagnosis is based on the window of opportunity that exists in the first few months of life when relatively simple treatment may be very effective. Methods range from wearing double nappies to splint therapy and corrective surgery. In general the later the diagnosis is made the harder the treatment will be, leading to greater risk of complications and a higher chance of failure [5]. There is a real need for a method of diagnosis that is simple, cheap, safe and effective, and US arguably provides such a technique. Unfortunately, the practice of US screening for DDH has developed with no randomized control trials to judge its efficacy, and the only evidence is from observational studies, albeit with very large numbers of cases [6].

In early infancy plain films will not show the femoral head or much of the acetabulum as these structures are not ossified until later in the first year of life. Whenever reasonable, plain film examination should be deferred until 3 to 6 months of age when more structures are ossified. Radiographs will demonstrate malalignment of the hips and show anomalies of the pelvis and sacrum. The initial plain film examination should be performed without any gonad shielding as this normally overlaps parts of the pelvic ring and sacrum. Defects in these areas such as sacral agenesis may otherwise be masked. Subsequent examination should use the shields to minimize radiation dose. Despite these comments, subtle or even moderate degrees of acetabular dysplasia will not be seen on plain films, especially in early infancy when treatment is more effective.

CT and MRI would be effective ways of examining the cartilaginous parts of the hips and they would allow assessment of the three-dimensional shape of the acetabulum. However, the high radiation burden from CT and the need for anaesthesia or sedation for most infants undergoing MRI preclude these as practical screening methods. US is safe, relatively cheap and repeatable with no need to sedate the infant. Its disadvantages are that it is labour-intensive and it requires skill and specific training both to perform and interpret the images. Studies have shown great sensitivity for US and a number of national bodies now require routine US screening of infants for hip dysplasia. Others, including those of the United Kingdom, recommend that US is used only in infants at high risk of developing DDH (Table 1.1).

Table 1.1 Risk factors for DDH

Female (not a criterion commonly used in high-risk screening protocols)
First degree relative with hip dysplasia
Premature birth
Breech presentation
Other congenital limb defects
Spinal defects

1.2.2.1 US Methods

1.2.2.1.1 Morphology

The method pioneered and developed by Reinhart Graf in Austria has gained the widest acceptance [7]. The infant is examined shortly after birth or at least in the first 6 weeks. The infant is laid in a foam-lined trough in the lateral decubitus position. The knee and hip of the uppermost side are flexed. The US probe is placed in a true coronal plane over the hip and the angle adjusted to give an image that shows the maximum depth of the acetabulum (Fig. 1.2).

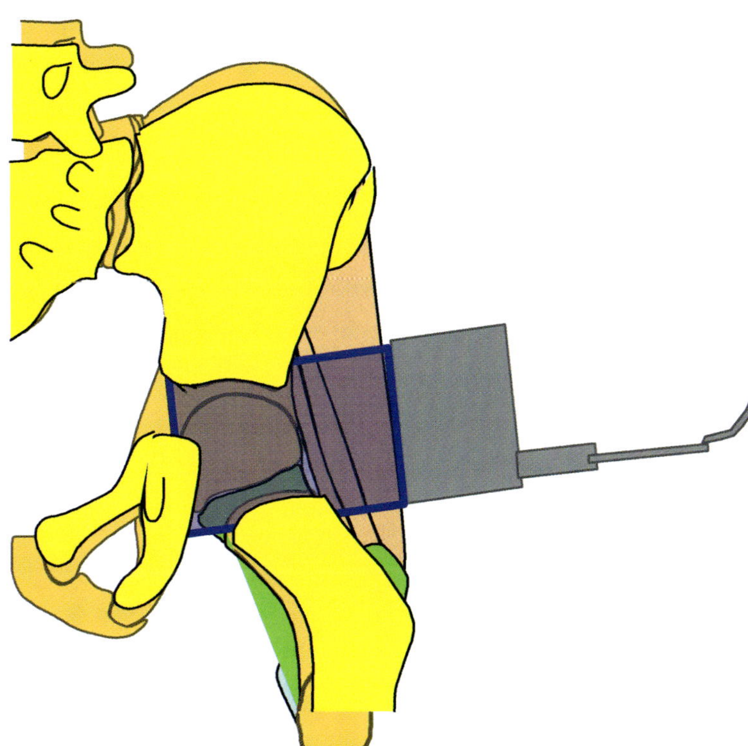

Fig. 1.2 US examination of the hip images the cartilaginous structures that are invisible on plain films in a coronal plane.

Care must be taken not to place the probe at an oblique angle to the coronal plane as the hip may be made to look erroneously deep or shallow. The need for a precise plane of imaging is a critical issue that demands training and audit of the technique. Measurements are made from the US image to assess the amount of the bony and cartilaginous cover of the femoral head by the acetabulum either using angles or the Morin (Terjesen) method in which the proportion of femoral head lying within the cavity is measured (Fig. 1.3) [8, 9], or the Graf technique (Fig. 1.4). Hips that are shallow in comparison to the normal population are reassessed at an interval of 1 or 2 weeks and if there is failure to develop normal acetabular cover then splint therapy is commenced. Immediate therapy is started without a follow-up study when the child has already reached an age where the opportunity to treat would be lost.

1.2.2.1.2
Dynamic Examination

Whilst there is some evidence that treatment may be based solely on the shape of the acetabulum, others argue that subluxation is a dynamic process and using the real-time capabilities of US it is possible to detect abnormal movement predicting dysplasia with perhaps greater sensitivity. The methods used vary but in general they are modifications of the stress tests of Ortolani and Barlow combined with US examination [10]. Gentle but firm pressure is placed on the upper part of the leg as if to subluxate the hip in a posterior and/or lateral direction. Movements of as little as 1 mm may be detected. How much movement is normal is contentious but some argue that over 2 mm of displacement on light stress is significant and requires treatment. It is probably wise to use both static and dynamic assessment in each case.

1.2.3
Role of Imaging in Treatment

The rate of splint therapy varies with individual practices and is said to be higher in those medical environments where strict conformity with treatment for abnormal US grading of acetabular dysplasia is applied.

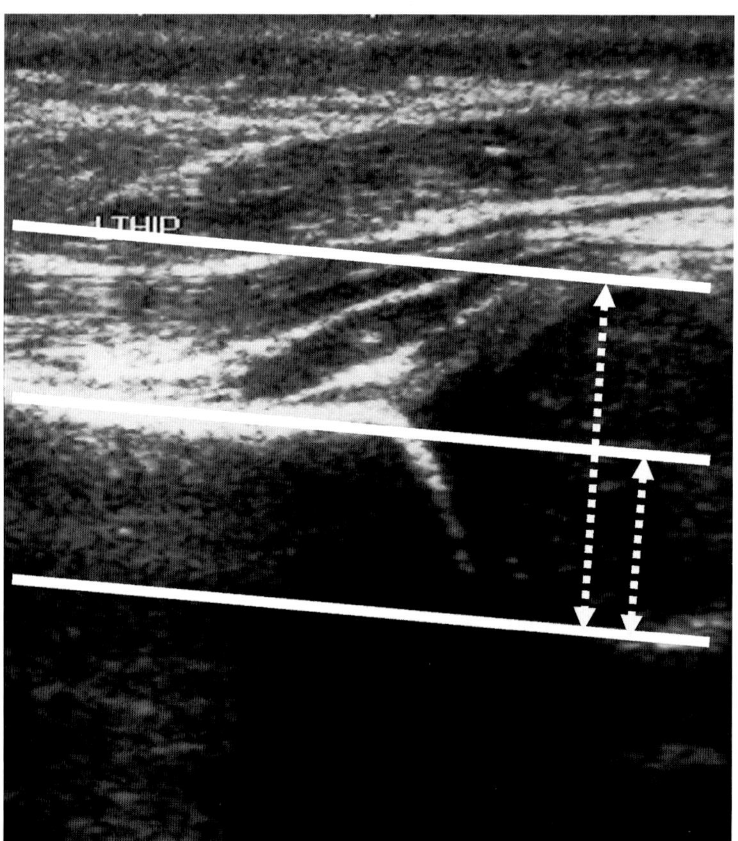

Fig. 1.3 US of the hip in the coronal plane with lines drawn to measure the amount of the head confined within the acetabulum (Morin/Terjesen method). A ratio of the overall width of the head is used as reference.

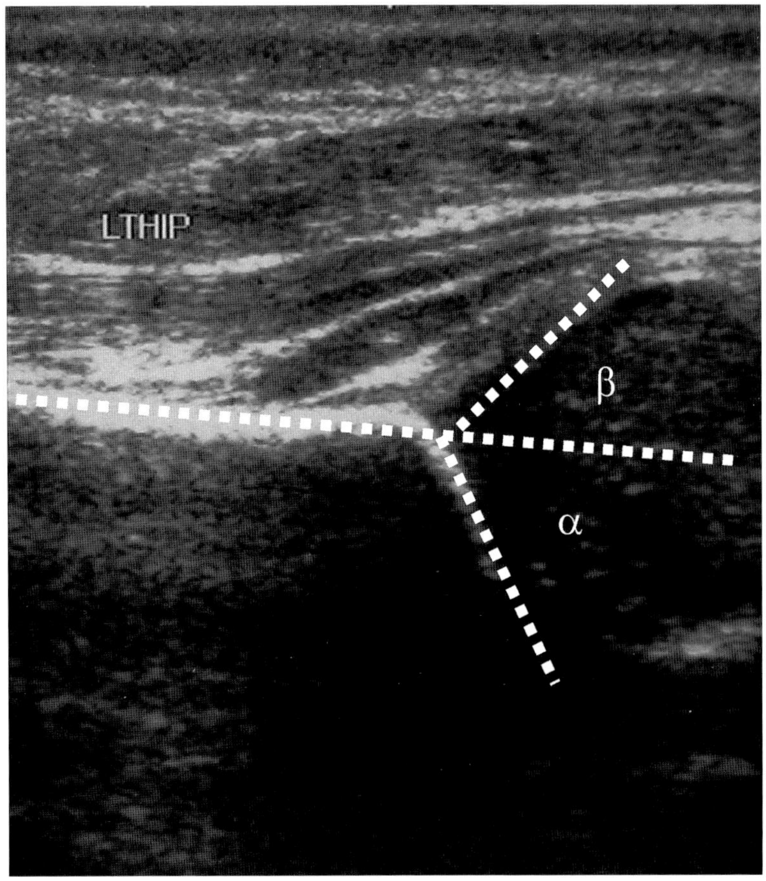

Fig. 1.4 US of the hip in the coronal plane with lines drawn to measure the Graf angles. A table of measurements is used to classify the shape of the hip [7].

Alternatively, it is argued that US screening may allow safe reduction in the numbers treated [11, 12].

Once an abnormal hip has been detected (Fig. 1.5) and treatment established there is a need to follow progress both of the shape of the acetabulum and the maturation of the bone. Over-aggressive manipulation and splint therapy may damage the growing epiphysis which will lead to deformity and delay in ossification. The latter is seen best on plain films or MRI. A reasonable approach is to repeat the US examination at follow-up appointments every 2 to 4 weeks during splint therapy [13] and then to perform a plain radiograph at the end of treatment or at 3 months of age (Fig. 1.6) [14]. Delay in ossification of the shallow side is expected but osteonecrosis will show much more severe retardation and then fragmentation. If there is doubt an MR study with coronal and axial T1- and T2-weighted images will detect or exclude femoral head necrosis.

When surgery is required to relocate a dislocated hip then imaging with an axial cross-section technique (CT or MRI) is important to ensure correct reduction [15–17]. Frontal view plain films may easily lead to posterior dislocation being overlooked. Lateral plain films are usually uninterpretable in a child with the hips in a plaster spica. The child is usually sedated and quiet immediately after surgery and the limbs are held in a cast; it is therefore relatively simple to acquire cross-sectional images. MR is the preferred technique to avoid radiation, although CT is equally effective (Fig. 1.7).

Planning of corrective osteotomies will require careful imaging. A combination of plain films, CT with thin low-dose sections and reconstruction, and MRI may be required [18]. Measurements may be taken from the workstation. Surface 3D reconstruction images are sometimes an aid to the surgeon. Newer software algorithms that give semitransparent images from multislice CT are especially useful as they mimic plane films and are better appreciated by those undertaking surgery.

1.2.4 Potential Developments

US examination is playing a greater role in the monitoring of suspect dysplastic hips [19] and will

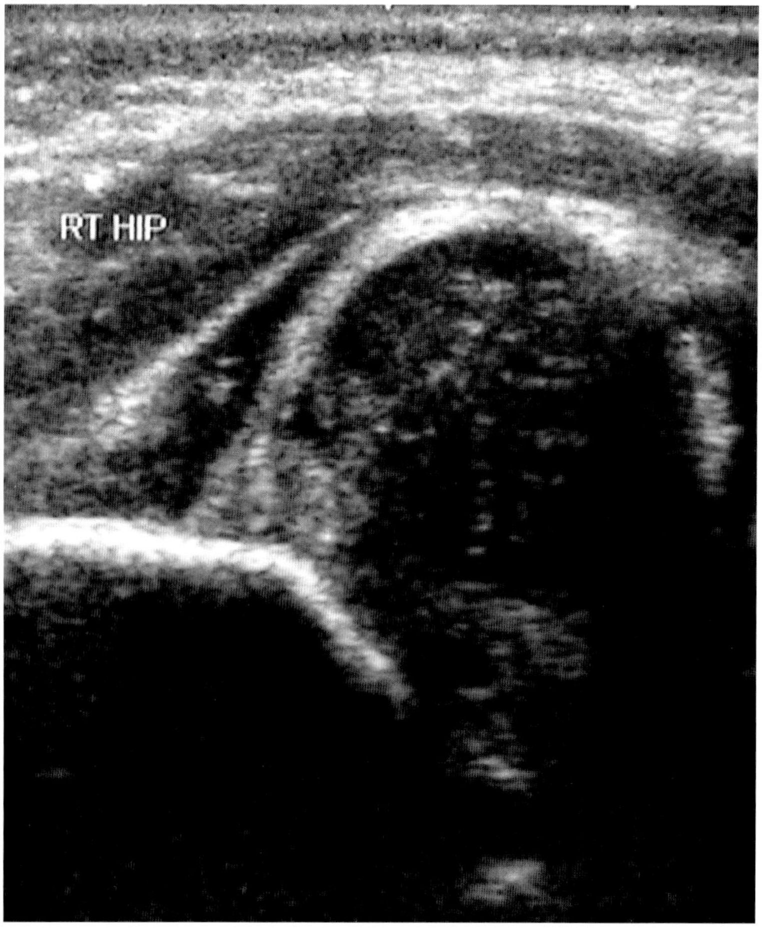

Fig. 1.5 US of a hip that is severely subluxed and almost dislocated. The "egg" of the femoral head is not sitting in the "spoon" of the acetabulum. The acetabular cartilage labrum is echogenic (bright), a sign seen when the tissue is stressed mechanically.

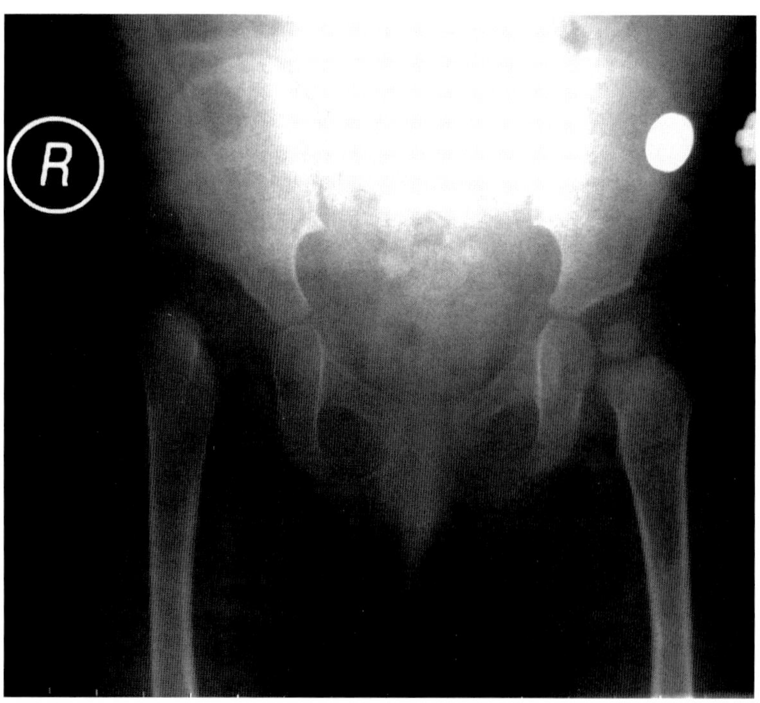

Fig. 1.6 The plain film appearances of the infant with hip subluxation seen at 3 months of age. The right femoral capital epiphysis has not ossified and the femur is aligned in a shortened and laterally placed position. The acetabulum is very shallow.

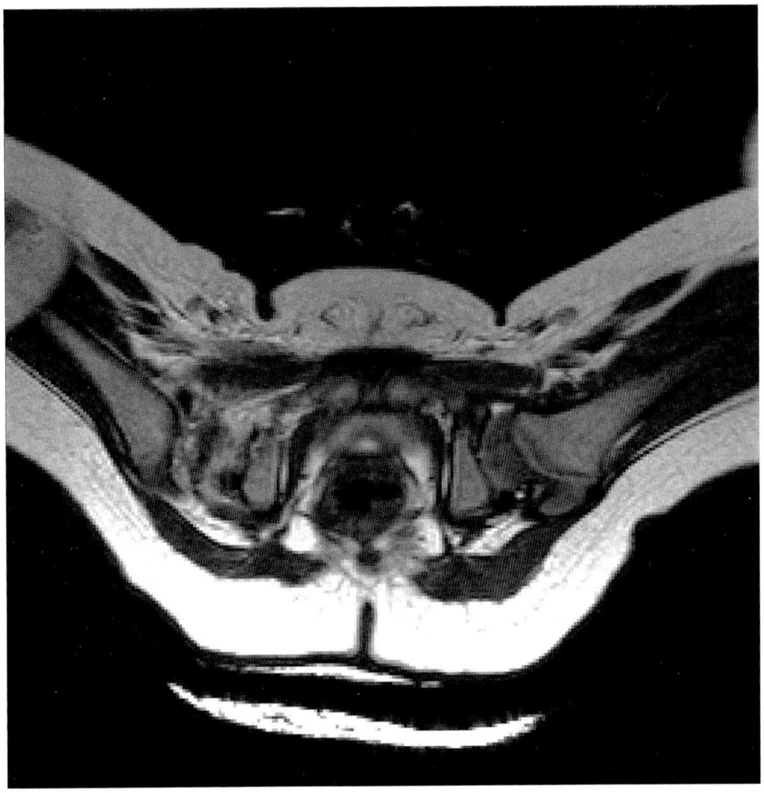

Fig. 1.7 Axial MRI immediately after surgical reduction of a dislocated hip with the pelvis in a plaster spica. The right femoral head is small and the acetabulum shallow but they are now properly aligned.

increasingly be used to determine the type and duration of treatment [20]. It is likely that our understanding of how and when to treat will advance as we use US to study outcome of therapy.

One area of contention is whether early treatment by splint therapy is effective. Large numbers of infants have been the subject of routine US screening and US-guided therapy in central European countries. Early data suggest that the incidence of late presentation dislocation of the hip may be much lower if not abolished [21, 22]. It will be interesting to see what happens to the rates of hip replacement in adults in the same population.

Doppler US or MRI with intravenous contrast agents has been advocated as a means of predicting osteonecrosis of the treated hip. Technically these are difficult examinations and these methods have not gained wide acceptance. It might be argued that once the damage to the vascular supply has occurred there is little that can be done to reverse the process, and the treatment will be salvage of what remains of the femoral head when the repair processes are complete.

Universal screening of all infants for DDH using US may seem a sensible approach but there is no consensus that this is reasonable at present [7, 23–28]. Not least is the doubt that splint treatment is necessary in all abnormal cases [29]. National policies on screening will in part reflect these awaited outcome studies but they may also be influenced by resources and health-care funding [30, 31]. The research will have to stand up to strict scrutiny before governments are likely to release the substantial funds required to establish universal US screening for DDH [32].

1.3 Focal Defects

1.3.1 Clinical Background

Apart from systemic disorders or syndromes there are infrequent cases of congenital limb deficiency or malformation. Thalidomide-associated phocomelia is the best known of this type of lesion (Fig. 1.8). Sporadic cases of unknown cause are the most frequent now that greater care is taken over prescribing any drugs during pregnancy. Focal defects include missing bones, absent joints, single forearm or lower

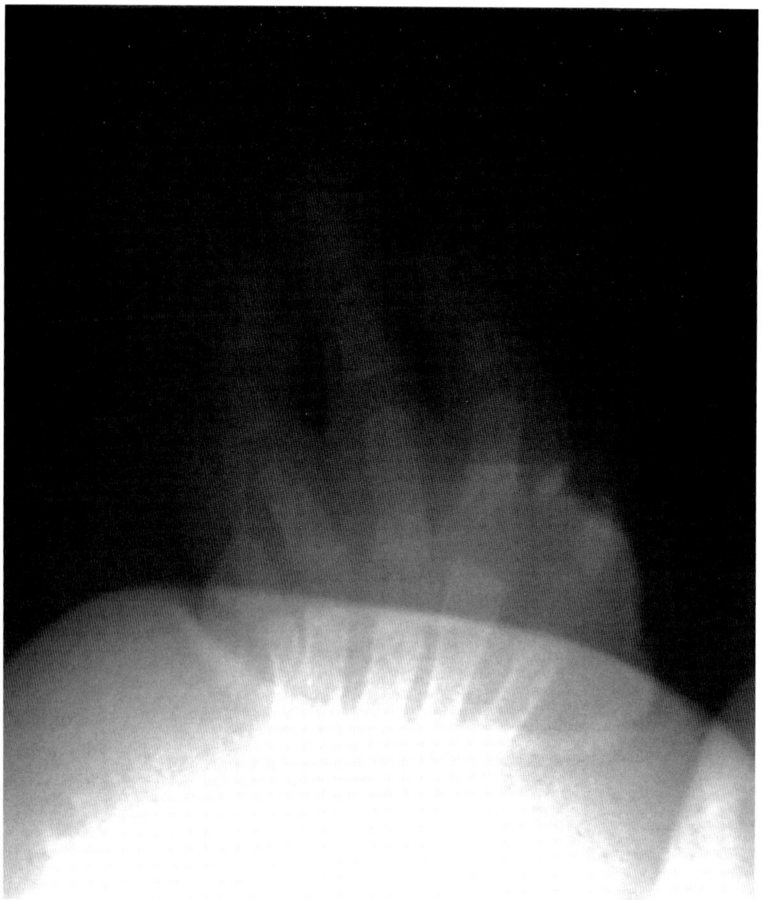

Fig. 1.8 The hand and vestigial upper limb of a child with phocomelia.

leg bones, and absence of a segment in a dermatomal pattern. There are sometimes associated abnormalities of other systems, e.g. Holt-Oram syndrome where radial deficiency in the forearm is associated with a cardiac lesion.

Defects of limb formation are now often recognized during pregnancy particularly at the 20-week "anomaly screening" examination [33–35]. It is common for the paediatric orthopaedic surgeon to be asked for advice on how such lesions might be treated by parents anticipating the need of their unborn child.

1.3.2 Role of Imaging

Imaging will be required to define the extent of the defect, predict progressive deformity that may occur during maturation and to plan surgical correction. In general the key is to define the anatomy as well as possible. Technically this is often very difficult. The infants are small and they move. The bone is not yet ossified and the structures involved are very abnormal in shape. A combination of imaging will be required.

Plain films are a prerequisite. They should be taken in planes as close to frontal and lateral as possible. Complex projections tend to confuse. MRI is very effective but the best surface coils and thinnest sections should be used. Conforming to true sagittal, coronal and axial planes will help. Conventional spin echo images are probably the easiest to interpret. Cortical bone will be of low signal on all sequences and difficult to see. Cartilage gives high signal on T2-weighted images. In the immature skeleton it is difficult to differentiate unossified cartilage from adjacent soft tissue. In a deformed limb the pattern and age of ossification is variable and unpredictable. A combination of CT and MR is useful as the bones are much better seen on CT and the cartilage is easiest to discriminate on T2-weighted images. US is very effective in showing unossified cartilage and the dynamic element allows the examiner to bend joints and demonstrate whether there is an intact joint or potential joint in an unossified cartilage block. US is most productive if performed after plain film and cross-sectional examinations. The examiner should have all previous imaging to hand before the US study. Rarely, contrast agents may be needed to demonstrate joint spaces; these may be introduced by needles guided by US and then imaged by fluoroscopy.

1.3.3 Potential Developments

Improved resolution of MR and US equipment will be invaluable in assessing these complex cases. The optimum timing for surgery and therefore imaging is not always clear and as experience increases this question may be answered.

1.4 Talipes Equinovarus

1.4.1 Clinical Background

Club foot is a condition of unknown cause, although it has been noted that the incidence is increased fourfold after amniocentesis. In some cases there is an association with a neurological defect, but there are also genetic and perhaps vascular factors [36, 37]. It usually presents at birth but the condition is increasingly being recognized at prenatal anomaly screening by US [38, 39]. There are several classification systems but none is linked to management protocols [40]. Treatment has been little changed for some time. Manipulation, splinting and often surgical soft tissue release are employed. For late problems, osteotomy and fusion are occasionally required [41].

1.4.2 Role of Imaging

Imaging is now often used to make an intrauterine diagnosis. For postnatal assessment some use MRI to assess the bony anatomy but the structures are very small and infants often will require anaesthesia for effective examination. Ossification of the hind foot bones is minimal in the infant where surgery is first considered. For this reason CT has little to offer, but plain films will help to clarify the overall alignment of the major bones. Plain radiographs are taken with an assistant holding a wooden block against the foot to achieve a "standing" position of the foot. The alignment of the hind foot is most important; MR studies have shown abnormal rotation and equinus of the calcaneus [42, 43]. The axis of the talus should align with the first metatarsal and the axis of the calcaneus should align with the fourth or fifth metatarsal (Fig. 1.9a). On a "standing" lateral view the talus should align with the first metatarsal whilst the calcaneus should make an angle of 10–30° with the talus and align with the first metatarsal (Fig. 1.9b). The observer should first judge the alignment of the hind foot as varus, valgus or normal. Then the relative position of the forefoot on the frontal (a.p.) view may be assessed. Hind foot valgus usually leads to a compensatory forefoot varus. The talus may be normally aligned or in a vertical position. The most common malalignment of the calcaneus is into "equinus" position. Named after the position of the horse's calcaneus, this implies an abnormal vertical alignment of the calcaneus with an excessively high arch to the mid-foot. US has the advantage of being dynamic and will assess the soft tissues in all but the most agitated of children [44]. It can also demonstrate the position of unossified bones [45]. As the aim of imaging is to define the abnormalities to allow planning of surgery, there must be close collaboration and understanding between the ultrasonographer and the surgeon. For this reason MR is probably the most useful technique [46].

MR imaging may be technically demanding as the deformity makes standard planes difficult to identify and reproduce. It is often easiest to strap the foot to a plastic or wooden splint to achieve as close to normal alignment as possible, this being equivalent to the walking position. The three conventional planes (coronal, sagittal and axial) are then used with sequences designed to contrast cartilage, muscle and tendon. T2-weighted fast spin echo is the most useful. Attention should be paid the number and alignment of the hind foot bones. Tibialis posterior tendon tension is often implicated and it is helpful to identify this tendon. Despite the potential for demonstrating the static anatomy, many surgeons will rely on clinical examination and the response to manipulation under anaesthesia for their diagnosis, classification and assessment. Postoperative imaging is probably best achieved with MRI [47, 48] when the position of bone, unossified cartilage and tendons may be studied.

There are links between lower limb deformity and spinal lesions so that careful clinical review of the spine with consideration of specific imaging is important in all children with foot deformities [49].

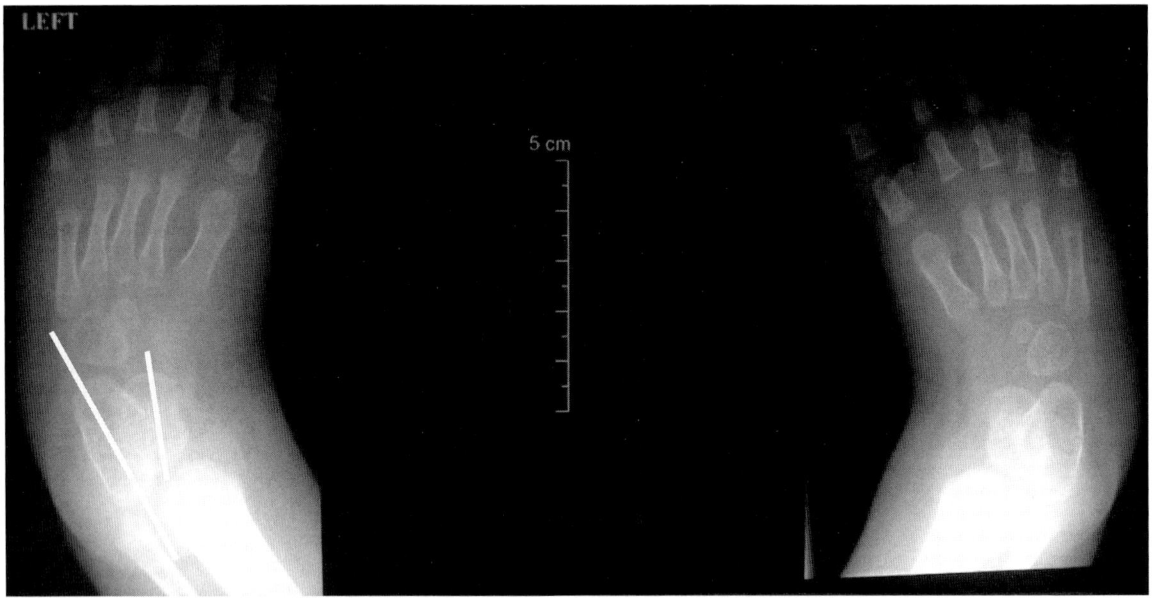

a

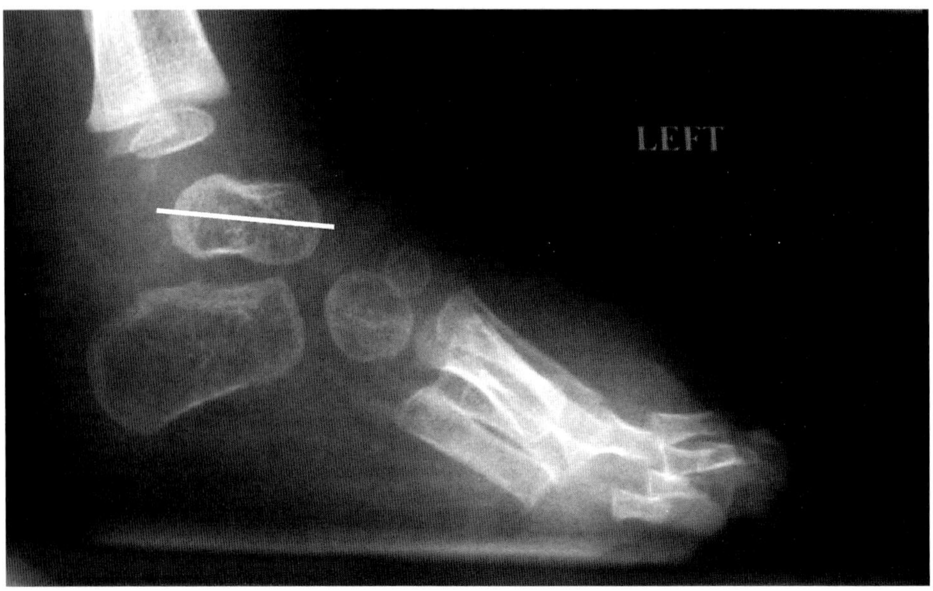

b

Fig. 1.9a,b. **a** AP; **b** lateral. Bilateral talipes equinovarus. Note that the axes of the calcaneus and the talus do not align respectively with the fourth/fifth metatarsals and the first metatarsal on the AP view. The talus does not align with the first metatarsal on the lateral view.

1.4.3 Potential Developments

Prenatal diagnosis will lead to prompt treatment. It may be that more effective management results. It seems to us that imaging is not being exploited effectively in the management decision-making, and there is a need for prospective studies using both MR and US. US has the potential to assess tethering and limitation of motion.

1.5 Neural Tube Defects

1.5.1 Clinical Background

Incomplete closure and errors in development of the neural tube in utero lead to the common clinical syndromes of spina bifida, myelomeningocele and secondary hydrocephalus. There is now a considerable expertise in the prenatal diagnosis of these lesions by US and this subject is dealt with in detail in many texts. As a result there is the option of termination of pregnancy with a reduction in the number of children born with these abnormalities. The most common presentation to imaging departments is now for the assessment of infants who have a sacral dimple or tuft of hair at the base of the spine.

Older children who have spinal column abnormalities including hemivertebrae, butterfly vertebrae, spinal cord tethering, diastematomyelia and syringomyelia may present with a deteriorating scoliosis. The management is often surgical with repair or release of tethered structures and instrumentation and osteotomy for the bony deformity.

1.5.2 Role of Imaging

Prenatal imaging is particularly important in allowing parents to made decisions regarding the continuance of pregnancy. US has significant advantages in accuracy over MRI, although both may be required in borderline or complex cases [50–52]. For open neural tube defects, closed myelomeningocele and cranial abnormalities MRI is the technique of choice [53]. This topic is dealt with in neuroradiological texts [54].

There are a number of disorders where the neural tube is intact but the bony architecture of the spine is abnormal. Children and adolescents who present with a lordoscoliosis or a kyphoscoliosis may be divided into those who have a congenital lesion (Fig. 1.10) such as a hemivertebra or spinal cord tethering and those who have a progressive structural change with no vertebral anomalies (idiopathic scoliosis and idiopathic kyphosis). Some adolescents may show endplate abnormalities that were not present in infancy; these include Scheuermann's disease and several skeletal dysplasias.

The imaging of vertebral column abnormalities has several goals:

1. To identify vertebral defects that might lead to progressive deformity
2. To identify neural tissue lesions that may damage the spinal cord function as the child matures
3. To measure the degree of deformity
4. To follow the progress of the disease and judge response to treatment
5. To plan surgery
6. To check for complications of surgery.

Techniques that are available are:

Plain films:

- Show vertebral defects
  - Hemivertebrae (Fig. 1.10)
  - Butterfly vertebrae (Fig. 1.11)
  - Wedged vertebrae
  - Fused (block) vertebrae
  - Endplate irregularity
- Show the overall alignment if taken whilst the child is standing
  - Require long films or detectors
  - Measurements are affected significantly by minor changes in projection
- Rotational deformities are difficult to measure and compare between examinations
- Mass neural lesions
  - Spinal cord tethering
  - Lipoma of the cord
  - Closed neural tube defects
  - Diastematomyelia (split cord)
  - Cord tumours
- Substantial radiation dose in young people
  - Limits repeat examination
- Films taken bending will show correctable (secondary) curves

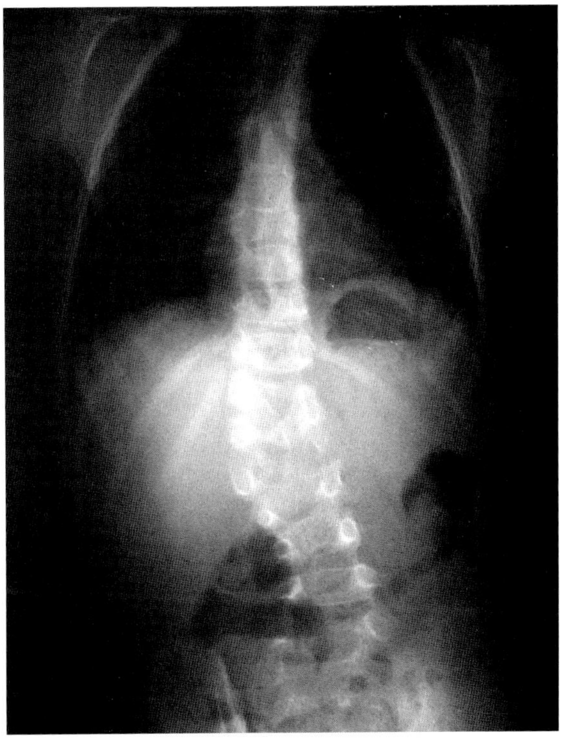

Fig. 1.10 Scoliosis with a short curve and vertebral anomalies. Two pedicles are missing on the left.

- "Cobb" angle measurement
  - Take the endplates of the vertebrae above and below the lesion that show the maximum angulation; measure the angle between these two endplates
  - Be aware that minor rotation in subsequent films will lead to a different result

Back shape photographic methods (photogrammetry):

- No radiation and easy to perform
  - Use projected light to image the shape of the back
  - Require the young person to undress
- Needs special equipment
  - Often bespoke and difficult to replace
- Addresses the commonest complaint—cosmetic deformity of the chest
- Does not show the underlying abnormalities
- Easy repletion and good reproducibility
- Allows for rotation in calculation of spinal curvature and chest wall deformity
- Measures the size of the chest wall "hump"

MRI:

- No ionizing radiation
- Shows all the bone and cord anomalies
- Requires a careful and complex series of sequences for example:
  - Cervical sagittal T2 fast spin echo
    - Foramen magnum defects
    - Chiari malformations (cerebellar tonsil herniation and fused vertebrae)
    - Syringomyelia
  - Thoracolumbar coronal T1 spin echo Scoliosis
    - Some vertebral anomalies especially hemivertebrae and butterfly vertebrae
    - Demonstrates kidneys (renal lesions are a common association with congenital spine deformity)
  - Thoracolumbar sagittal T2 fast spin echo
    - Spinal cord tethering
    - Fused vertebrae
    - Meningocele
    - Lipoma of the cord
    - Cord tumours
  - Thoracolumbar axial T2* gradient echo (wide coverage)
    - Split cord (may be missed on coronal and sagittal images)
    - Diastematomyelia (Fig. 1.12)
    - Meningocele
- Young children may need to be sedated
- Cannot be performed standing (except in very uncommon standing MR units)

US:

- No ionizing radiation
- Limited to soft tissue changes
- Spinal cord masked by the vertebral arch More useful in infants
- Shows CSF pulsation
- Sedation not required
- Effective in excluding cord tethering and neural tube defects in infancy

Myelography (with or without CT):

- An outdated technique replaced by MR
- Rarely needed if MRI is contraindicated, e.g. cranial surgical clips
- Invasive and difficult
- Cannot show internal lesions of the cord
- Radiation dose substantial as a wide area of examination is important

CT:

- A useful adjunct to MR in complex bone deformity
- Requires complex multiplane reconstruction
- Best viewed on a workstation

Infants with tufts of hair at the base of the spine and sacral dimples are most often normal. The role of imaging is to exclude meningoceles, spinal cord tethering and large bony neural arch defects. Care should be taken not to alarm the parents and family when there is an isolated bony arch defect as these are very common in the normal asymptomatic adult population. In the newborn infant ossification of the cartilage bony arch progresses from the region of the pedicles and it is easy to look at the partial ossification margins and regard them as abnormal. The most effective imaging method is US [55–57]. It is fast and accurate. The infant may be examined whilst held against the parent's chest. A linear array high-resolution probe is required and extended view imaging assists (Fig. 1.13). The examiner should identify the conus medullaris which should have its tip at around the first lumbar vertebra (Fig. 1.14). The neural arch is best seen on axial images (Figs. 1.15, 1.16). The conus moves with respiration. Tethering will reduce the movement and pull the conus lower down the canal. Fat is echogenic (white) and a lipoma of the filum will be clearly differentiated from the echo-free (black) cerebrospinal fluid (CSF). Meningoceles

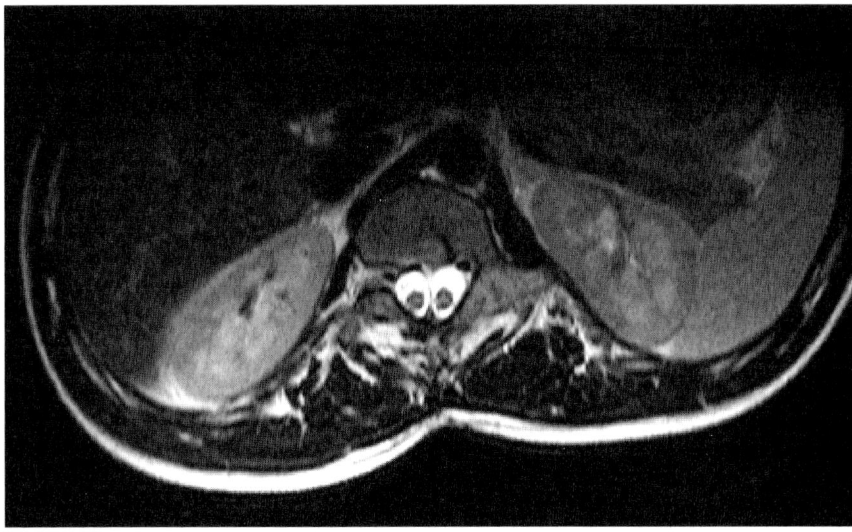

Fig. 1.12 MRI of a diastematomyelia.

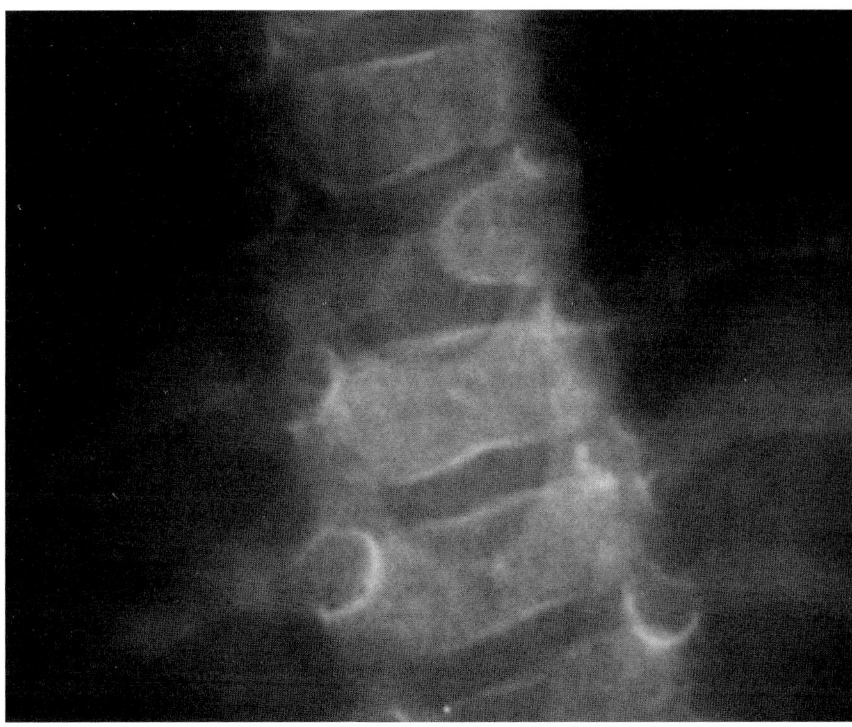

Fig. 1.11 A butterfly vertebra.

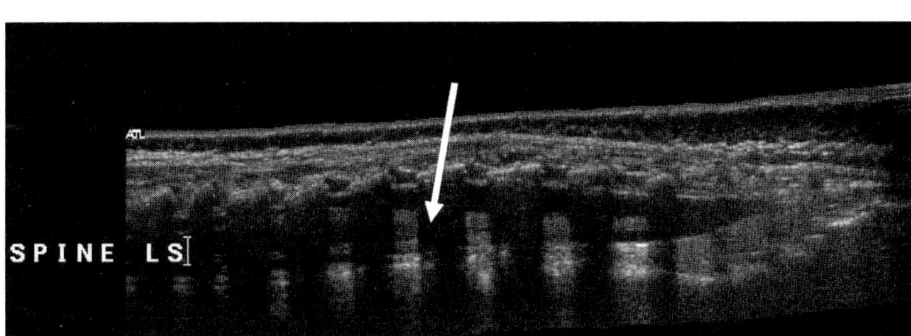

Fig. 1.13 Sagittal extended-view US image of a normal cord. The conus finishes at the arrow.

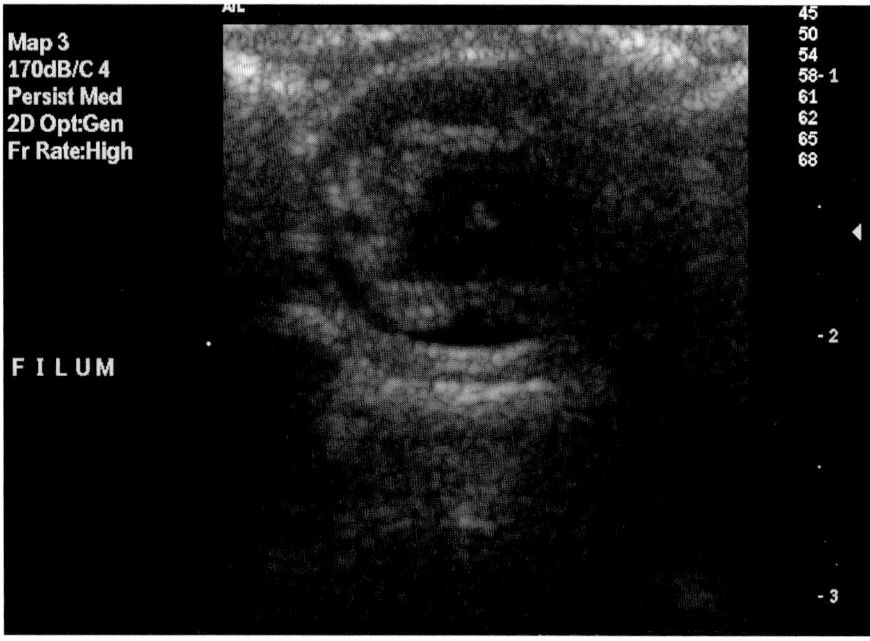

Fig. 1.14 Axial US image of a normal filum terminale.

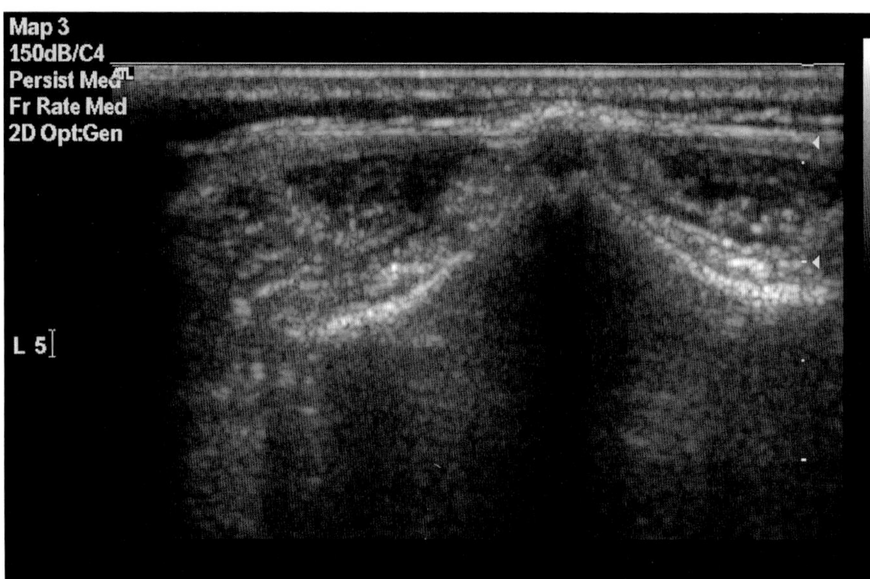

Fig. 1.15 Axial US image of an intact neural arch.

will contain CSF and communications will be identified by the neck or isthmus. Their communication with the central canal will be demonstrable by pulsation of CSF.

MR should be used in doubtful or complex cases [58, 59] (Figs. 1.17, 1.18). MRI will be needed when abnormalities are found and treatment is being considered. It provides a better "road map" for the surgeon [60].

We suggest the following protocols:

- Suspected neural tube defect in infancy: US; if abnormal then MRI
- Scoliosis: plain film standing; if smooth curve then treat; if short radius curve, vertebral defects, pain or neurological symptoms then MRI
- MRI difficult to interpret: CT
- MRI contraindicated: CT myelography
- Conservative treatment follow-up: photogrammetry
- Surgical follow-up: plain films standing; if difficult to interpret then CT

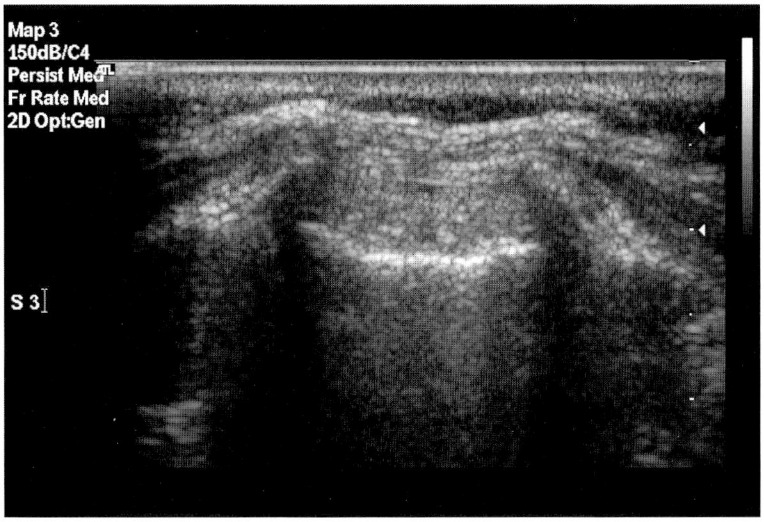

Fig. 1.16 Axial US image of a bifid neural arch.

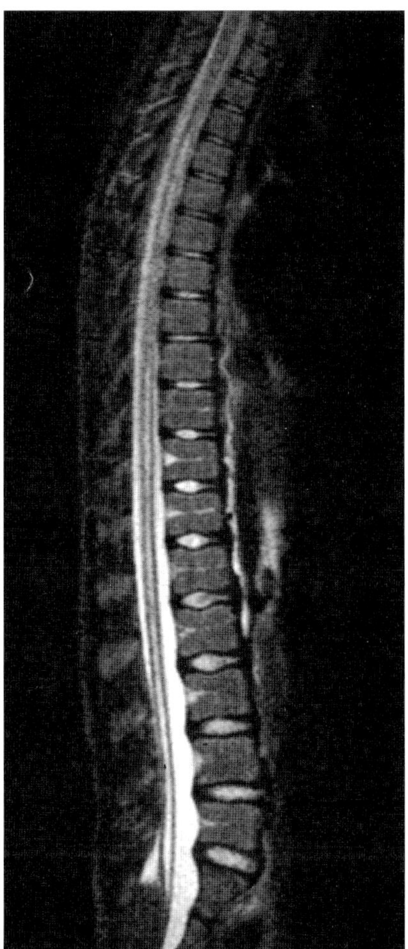

Fig. 1.17 Sagittal fat-suppressed T2-weighted MR image of a child with a tethered cord and syringomyelia.

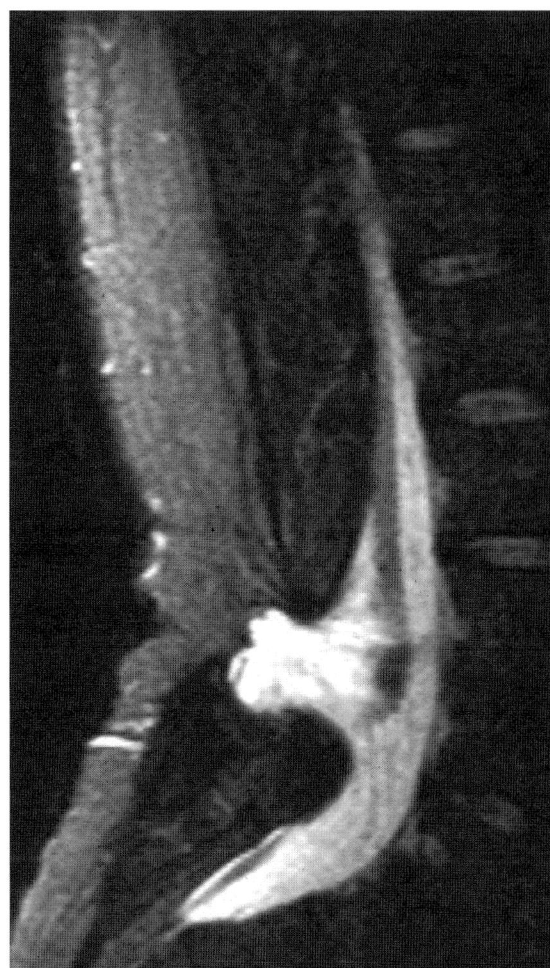

Fig. 1.18 Sagittal fat-suppressed T2-weighted MR image of a child with a myelomeningocele.

1.5.3 Potential Developments

US assessment of dimples and hair tufts is only available in a limited number of centres. Training and experience will expand its use.

References and Further Reading

1. Shefelbine SJ, Carter DR (2004) Mechanobiological predictions of growth front morphology in developmental hip dysplasia. J Orthop Res 22(2):346–352
2. Bialik V, Bialik GM, Blazer S, et al (1999) Developmental dysplasia of the hip: a new approach to incidence. Pediatrics 103(1):93–99
3. Kobayashi S, Saito N, Nawata M, et al (2004) Total hip arthroplasty with bulk femoral head autograft for acetabular reconstruction in DDH. Surgical technique. J Bone Joint Surg Am 86 [Suppl 1]:11–17
4. Sahin F, Akturk A, Beyazova U, et al (2004) Screening for developmental dysplasia of the hip: results of a 7-year follow-up study. Pediatr Int 46(2):162–166
5. Guille JT, Pizzutillo PD, MacEwen GD (2000) Development dysplasia of the hip from birth to six months. J Am Acad Orthop Surg 8(4):232–242
6. Puhan MA, Woolacott N, Kleijnen J, et al (2003) Observational studies on ultrasound screening for developmental dysplasia of the hip in newborns—a systematic review. Ultraschall Med 24(6):377–382
7. Graf R (2002) Profile of radiologic-orthopedic requirements in pediatric hip dysplasia, coxitis and epiphyseolysis capitis femoris (in German). Radiologe 42(6):467–473
8. Czubak J, Kotwicki T, Ponitek T, et al (1998) Ultrasound measurements of the newborn hip. Comparison of two methods in 657 newborns. Acta Orthop Scand 69(1):21–24
9. Terjesen T (1996) Ultrasound as the primary imaging method in the diagnosis of hip dysplasia in children aged <2 years. J Pediatr Orthop B 5(2):123–128
10. Harcke HT, Grissom LE (1999) Pediatric hip sonography. Diagnosis and differential diagnosis. Radiol Clin North Am 37(4):787–796
11. Riboni G, Bellini A, Serantoni S, et al (2003) Ultrasound screening for developmental dysplasia of the hip. Pediatr Radiol 33(7):475–481
12. Sampath JS, Deakin S, Paton RW (2003) Splintage in developmental dysplasia of the hip: how low can we go? J Pediatr Orthop 23(3):352–355
13. Malkawi H (1998) Sonographic monitoring of the treatment of developmental disturbances of the hip by the Pavlik harness. J Pediatr Orthop B 7(2):144–149
14. Ucar DH, Isiklar ZU, Kandemir U, et al (2004) Treatment of developmental dysplasia of the hip with Pavlik harness: prospective study in Graf type IIc or more severe hips. J Pediatr Orthop B 13(2):70–74
15. Laor T, Roy DR, Mehlman CT (2000) Limited magnetic resonance imaging examination after surgical reduction of developmental dysplasia of the hip. J Pediatr Orthop 20(5):572–574
16. Westhoff B, Wild A, Seller K, et al (2003) Magnetic resonance imaging after reduction for congenital dislocation of the hip. Arch Orthop Trauma Surg 123(6):289–292
17. McNally EG, Tasker A, Benson MK (1997) MRI after operative reduction for developmental dysplasia of the hip. J Bone Joint Surg Br 79(5):724–726
18. Kim SS, Frick SL, Wenger DR (1999) Anteversion of the acetabulum in developmental dysplasia of the hip: analysis with computed tomography. J Pediatr Orthop 19(4):438–442
19. Gerscovich EO (1997) A radiologist's guide to the imaging in the diagnosis and treatment of developmental dysplasia of the hip. II. Ultrasonography: anatomy, technique, acetabular angle measurements, acetabular coverage of femoral head, acetabular cartilage thickness, three-dimensional technique, screening of newborns, study of older children. Skeletal Radiol 26(8):447–456
20. Hedequist D, Kasser J, Emans J (2003) Use of an abduction brace for developmental dysplasia of the hip after failure of Pavlik harness use. J Pediatr Orthop 23(2):175–177
21. Weitzel D (2002) [Ultrasound screening of the infant hip]. Radiologe 42(8):637–645
22. Wirth T, Stratmann L, Hinrichs F (2004) Evolution of late presenting developmental dysplasia of the hip and associated surgical procedures after 14 years of neonatal ultrasound screening. J Bone Joint Surg Br 86(4):585–589
23. Toma P, Valle M, Rossi U, et al (2001) Paediatric hip—ultrasound screening for developmental dysplasia of the hip: a review. Eur J Ultrasound 14(1):45–55
24. Karapinar L, Surenkok F, Ozturk H, et al (2002) The importance of predicted risk factors in developmental hip dysplasia: an ultrasonographic screening program (in Turkish). Acta Orthop Traumatol Turc 36(2):106–110
25. Rosenberg N, Bialik V (2002) The effectiveness of combined clinical-sonographic screening in the treatment of neonatal hip instability. Eur J Ultrasound 15(1–2):55–60
26. Zenios M, Wilson B, Galasko CS (2000) The effect of selective ultrasound screening on late presenting DDH. J Pediatr Orthop B 9(4):244–247
27. Kocher MS (2000) Ultrasonographic screening for developmental dysplasia of the hip: an epidemiologic analysis (Part I). Am J Orthop 29(12):929–933
28. Lewis K, Jones DA, Powell N (1999) Ultrasound and neonatal hip screening: the five-year results of a prospective study in high-risk babies. J Pediatr Orthop 19(6):760–762
29. Lorente Molto FJ, Gregori AM, Casas LM, et al (2002) Three-year prospective study of developmental dysplasia of the hip at birth: should all dislocated or dislocatable hips be treated? J Pediatr Orthop 22(5):613–621
30. Clegg J, Bache CE, Raut VV (1999) Financial justification for routine ultrasound screening of the neonatal hip. J Bone Joint Surg Br 81(5):852–857
31. Patel H (2001) Preventive health care. 2001 update: screening and management of developmental dysplasia of the hip in newborns. CMAJ 164(12):1669–1677
32. Eastwood DM (2003) Neonatal hip screening. Lancet 361(9357):595–597
33. Ryu JK, Cho JY, Choi JS (2003) Prenatal sonographic diagnosis of focal musculoskeletal anomalies. Korean J Radiol 4(4):243–251
34. Camera G, Dodero D, Parodi M, et al (1993) Antenatal ultrasonographic diagnosis of a proximal femoral focal deficiency. J Clin Ultrasound 21(7):475–479

35. Seow KM, Huang LW, Lin YH, et al (2004) Prenatal three-dimensional ultrasound diagnosis of a camptomelic dysplasia. Arch Gynecol Obstet 269(2):142–144
36. Kammoun F, Tanguy A, Boesplug-Tanguy O, et al (2004) Club feet with congenital perisylvian polymicrogyria possibly due to bifocal ischemic damage of the neuraxis in utero. Am J Med Genet 126A(2):191–196
37. Ng YT, Mancias P, Butler IJ (2002) Lumbar spinal stenosis causing congenital clubfoot. J Child Neurol 17(1):72–74
38. Mohammed NB, Biswas A (2002) Three-dimensional ultrasound in prenatal counselling of congenital talipes equinovarus. Int J Gynaecol Obstet 79(1):63–65
39. Keret D, Ezra E, Lokiec F, et al (2002) Efficacy of prenatal ultrasonography in confirmed club foot. J Bone Joint Surg Br 84(7):1015–1019
40. Roye DP Jr, Roye BD (2002) Idiopathic congenital talipes equinovarus. J Am Acad Orthop Surg 10(4):239–248
41. Roye BD, Hyman J, Roye DP Jr (2004) Congenital idiopathic talipes equinovarus. Pediatr Rev 25(4):124–130
42. Saito S, Hatori M, Kokubun S, et al (2004) Evaluation of calcaneal malposition by magnetic resonance imaging in the infantile clubfoot. J Pediatr Orthop B 13(2):99–102
43. Kamegaya M, Shinohara Y, Kuniyoshi K, et al (2001) MRI study of talonavicular alignment in club foot. J Bone Joint Surg Br 83(5):726–730
44. Hamel J (2002) Ultrasound diagnosis of congenital foot deformities (in German). Orthopade 31(3):326–327
45. Aurell Y, Johansson A, Hansson G, et al (2002) Ultrasound anatomy in the neonatal clubfoot. Eur Radiol 12(10):2509–2517
46. Cahuzac JP, Navascues J, Baunin C, et al (2002) Assessment of the position of the navicular by three-dimensional magnetic resonance imaging in infant foot deformities. J Pediatr Orthop B 11(2):134–138
47. Pirani S, Zeznik L, Hodges D (2001) Magnetic resonance imaging study of the congenital clubfoot treated with the Ponseti method. J Pediatr Orthop 21(6):719–726
48. Pekindil G, Aktas S, Saridogan K, et al (2001) Magnetic resonance imaging in follow-up of treated clubfoot during childhood. Eur J Radiol 37(2):123–129
49. Ward PJ, Clarke NM, Fairhurst JJ (1998) The role of magnetic resonance imaging in the investigation of spinal dysraphism in the child with lower limb abnormality. J Pediatr Orthop B 7(2):141–143
50. Blaicher W, Mittermayer C, Messerschmidt A, et al (2004) Fetal skeletal deformities—the diagnostic accuracy of prenatal ultrasonography and fetal magnetic resonance imaging. Ultraschall Med 25(3):195–199
51. Patel TR, Bannister CM, Thorne J (2003) A study of prenatal ultrasound and postnatal magnetic imaging in the diagnosis of central nervous system abnormalities. Eur J Pediatr Surg 13 [Suppl 1]:S18–22
52. Oi S (2003) Current status of prenatal management of fetal spina bifida in the world: worldwide cooperative survey on the medico-ethical issue. Childs Nerv Syst 19(7–8):596–599
53. Rossi A, Cama A, Piatelli G, et al (2004) Spinal dysraphism: MR imaging rationale. J Neuroradiol 31(1):3–24
54. Verity C, Firth H, Ffrench-Constant C (2003) Congenital abnormalities of the central nervous system. J Neurol Neurosurg Psychiatry 74 [Suppl 1]:i3–8
55. Hughes JA, De Bruyn R, Patel K, et al (2003) Evaluation of spinal ultrasound in spinal dysraphism. Clin Radiol 58(3):227–233
56. Dick EA, de Bruyn R (2003) Ultrasound of the spinal cord in children: its role. Eur Radiol 13(3):552–562
57. Dick EA, Patel K, Owens CM, et al (2002) Spinal ultrasound in infants. Br J Radiol 75(892):384–392
58. Allen RM, Sandquist MA, Piatt JH Jr, et al (2003) Ultrasonographic screening in infants with isolated spinal strawberry nevi. J Neurosurg Spine 98(3):247–250
59. Tortori-Donati P, Rossi A, Biancheri R, et al (2001) Magnetic resonance imaging of spinal dysraphism. Top Magn Reson Imaging 12(6):375–409
60. Khanna AJ, Wasserman BA, Sponseller PD (2003) Magnetic resonance imaging of the pediatric spine. J Am Acad Orthop Surg 11(4):248–259

2 Trauma and Sports-related Injuries

Philip J. O'Connor and Clare Groves

CONTENTS

2.1 General Principles 19
2.1.1 Biomechanics 19
2.1.2 Imaging 20
2.2 Acute Trauma 20
2.2.1 Acute Fracture Patterns 20
2.2.1.1 Diaphyseal and Metaphyseal Injuries 20
2.2.1.2 Physeal Injuries 21
2.2.1.3 Apophyseal Injuries 21
2.2.1.4 Acute Osteochondral Injuries 22
2.2.2 Foreign Bodies 24
2.2.3 Haematoma 25
2.2.4 Muscles and Tendons 26
2.2.4.1 Shoulder 26
2.3 Chronic Trauma 28
2.3.1 Chronic Fracture Patterns 28
2.4 The Osteochondroses 32
2.4.1 Osteochondritis Dissecans 32
2.4.2 Panner's Disease 33
2.4.3 Medial Epicondylitis (Little League Elbow) 34
2.5 Accessory Ossicles 34
2.6 Bursae 34
2.7 Summary 37
References and Further Reading 37

2.1 General Principles

Injury is the response of tissue to kinetic energy applied to the body. Damage may occur locally or distant from the site of trauma due to transmitted forces, and may be acute or chronic arising from repetitive strains. Chronic overuse injuries are particularly important in the young athlete. There are fundamental differences in the young skeleton and that of the mature adult, which lead to disparate patterns of injury from the same degree of force. In order to understand patterns of skeletal injury one first must understand their kinetic chain and the effect of force upon it. The aim of this chapter is to give the reader an understanding of the factors affecting the nature of skeletal injury with specific emphasis on the role of musculoskeletal ultrasound (US).

P. J. O'Connor, FRCR
C. Groves, FRCR
Department of Radiology, The General Infirmary at Leeds, Leeds, LS1 3EX, UK

2.1.1 Biomechanics

The kinetic chain is the functional unit that allows us to move the skeleton. The skeleton provides essential soft tissue support with joints determining the body's range of movement. Muscles and tendons provide the forces to actively move and control the skeleton while also serving as active stabilizers along with ligaments and capsule giving soft-tissue stability to joints. The nature of injury to these structures results from the application of force to these elements.

Any force if large enough will produce failure in the skeleton in a predictable way. The site of failure will usually be at the weakest point within the structure, this varies with the age of the patient and obviously differs depending on the forces applied. In the skeletally immature patient the kinetic chain differs from adults in that growth plates are present around joints and at apophyses (tendon bone junctions). A large acute force will usually result in bone injury at its weakest point. This is the junction between mature and growing bone, i.e. the epiphysis or apophysis [1]. Fractures in patients of this age are thus usually either apophyseal avulsions or Salter-Harris type injuries to the growth plate. Repetitive strain is a common mechanism for sports-related injury and occurs as a result of forces large enough to damage but not cause structural failure of a tissue. The insult is then reapplied cyclically (i.e. during training) before complete tissue healing occurs. With each cycle the tissue weakens until eventually the force applied is larger than the tissue tolerance and complete structural failure ensues. These forces are usually complex as a result of differing sports and patient biomechanics, although they will have either a predominantly passive compressive or active distractive nature.

Passive compressive forces result more in damage to osseous structures and are particularly seen in association with high impact cyclical injury (i.e. long-distance running on hard surfaces). In the skeletally immature patient injury again usually occurs at the site of growing bone. The very young and in those patients approaching maturity, failure can occur elsewhere in the kinetic chain. The diaphysis of long bones as in the very young can be the site of injury as the bone itself has differing mechanical properties making this the weakest point. In older patients fusing epiphysis similarly no longer represents the weakest point in the chain and compressive forces can result in stress injury to the diaphysis. Changes can be seen within joints and are normally seen in association with compressive or rotational (twisting and varus/valgus stress) forces. Within joints osteochondral injury occurs much more commonly than internal or ligamentous disruption except where there are pre-existing congenital variants such as discoid menisci in the knee.

Active forces are related to the contraction of the muscle tendon unit. Injury most commonly occurs in muscles crossing two joints as these are inherently subject to greater forces. Common examples of such muscles are the biceps in the upper limb or the gastrocnemii, hamstrings and the rectus femoris in the lower limb. In the musculoskeletally immature patient the apophysis represents the weakest point in this chain and is thus the most commonly injured site in cyclical injuries. As the patient approaches maturity an increase in incidence of musculotendinous junction injuries will become apparent as the apophyses begin to fuse.

In general the type of force and the age of the patient tend to determine the site at which that failure will occur within the musculoskeletal system, with the biomechanics of the individual determining the pattern of injury. For example, patients who are skeletally mature presenting with calf muscles tears. The nature of the force will be very similar in all patients—normal explosive contraction of the calf muscles. The musculotendinous junction is the point of structural failure with the biomechanics determining which muscle will fail. For example, some patients tear soleus rather than gastrocnemius and some patients tear the lateral rather than medial musculotendinous junction. The individual's biomechanics determine the pattern of injury with the site of failure determined by the nature of the force and the age of the patient.

2.1.2 Imaging

Management of paediatric trauma requires close communication between the clinician and the investigating radiologist. The clinical history is vital, since the mechanism will usually predict the likely injury. Appropriate imaging may then be requested. For example, suspected muscle or tendon rupture is best assessed with US, while stress fractures may be missed on plain film and require radionuclide scintigraphy. In the adolescent, osteochondral injuries are commonly encountered and these require cross-sectional imaging, usually with MR. Special consideration should be given to the young athlete who is more likely to suffer from chronic overuse syndromes. The patterns of injury may be predicted from the type of sport, with lower limb injuries often arising from football and basket ball, upper limb in baseball and swimming, and overuse injuries in swimming, gymnastics and throwing sports [2].

2.2 Acute Trauma

2.2.1 Acute Fracture Patterns

2.2.1.1 Diaphyseal and Metaphyseal Injuries

The biomechanical properties of growing bone may lead to incomplete, greenstick fractures, which are peculiar to children. Immature bone is more porous and less dense than adult bone due to increased vascular channels and a lower mineral content. Increased plasticity and elasticity of young bone means that it is more likely to bend or buckle than to snap. The periosteum is thicker, more elastic and less firmly bound to bone, so it will usually remain intact over an underlying fracture. Healing and remodelling is therefore more predictable than in adults and non-union is rare.

Rogers [1] classifies these injuries broadly as classic greenstick, torus and bowing fractures. The classic greenstick fracture arises from bending forces, which produce a complete break of the cortex on the tension side and plastic deformation of the opposite cortical border. The resulting fracture line may then extend at right angles to its medial extent, causing a longitudinal split in the shaft. Classic greenstick fractures are seen in the mid-shaft of the radius, ulna and clavicle. The torus fracture is produced by compressive forces, which cause the cortex to buckle, and occurs most commonly in the distal radius and ulna. Bowing or plastic fractures result from angulation and compressive forces producing a gradual curve across the length of the whole bone. These may be easily overlooked by the unwary, and occur most frequently in the distal radius and ulna.

Spiral fractures due to torsion may occur in the lower limbs where they have been termed toddler's fractures. Typically these are a non-displaced oblique fracture of the distal tibia but it is recognized that other bones may be involved, and most present between the ages of 1 and 3 years [3]. Toddler's fractures rarely involve the epiphyses, and are usually solitary injuries with little or no associated bruising. These are important discriminators when the possibility of child abuse is being considered.

2.2.1.2 Physeal Injuries

Damage to the physis may lead to temporary or permanent disturbance of growth. Growth plate injuries are usually classified according to the Salter-Harris system where the higher the value assigned to a fracture pattern the worse the prognosis. Most of these fractures are apparent on plain radiographs, which may demonstrate epiphyseal displacement. Other signs include widening of the physis, and loss of definition of the opposing surfaces of the epiphysis and metaphysis [4]. The peak incidence of growth-plate injury occurs during adolescence, perhaps due to increased exposure to high-energy trauma combined with weakening of the growth plate that occurs with puberty [5]. Epiphyseal injuries are commoner in males, probably due to greater exposure to trauma and sport, and to the relative delay in epiphyseal closure [1].

2.2.1.3 Apophyseal Injuries

An apophysis is a growth centre where a tendon attaches to bone. Its cartilaginous growth plate remains weaker than the attached musculotendinous unit until it fuses at the time of skeletal maturation (Table 2.1). Applying the principle of failure at the weakest link, it is predictable that a sudden forceful muscular contraction may cause apophyseal separation. Young athletes are prone to apophyseal avulsion, particularly in sports requiring sudden powerful acceleration or change of direction such as football, dance and gymnastics. Apophyseal injuries occur most commonly around the pelvis, with the most frequent sites being the anterior inferior iliac spine, anterior superior iliac spine, iliac crest and ischial tuberosity [6]. Acute avulsion injuries are seen on the plain radiograph as crescentic osseous fragments and may heal with abundant callus due to associated haematoma (Fig. 2.1) [7]. This may lead to unusual radiographic appearances, which can be mistaken for tumour. MR imaging may help to clarify the diagnosis and avoid a potentially misleading biopsy [8]. Note that healing fractures may exhibit cellular change that can be similar to that seen in aggressive tumours.

Table 2.1 Common sites of apophyseal avulsion in the hip and pelvis with the age of fusion and the responsible avulsing muscle group [6]

Apophyses	Fusion (years)	Related muscle group
Anterior inferior iliac spine	16–18	Quadriceps
Anterior superior iliac spine	21–25	Sartorius
Lesser trochanter	16–17	Iliopsoas
Greater trochanter	16–17	Gluteal
Ischial tuberosity	20–25	Hamstrings
Iliac crest	21–25	Abdominal obliques, latissimus dorsi

Clinically, it is often difficult to distinguish between a simple muscle strain and an apophyseal avulsion. The plain radiograph may be unhelpful if the ossification centre of the apophysis has not yet formed. However, in young athletes the correct diagnosis is necessary to establish the appropriate treatment and rehabilitation programme. Lazovic et al. [9] described the use of US in 243 cases of suspected apophyseal avulsion. They found it to be more sensitive than plain radiography, with the advantage that US allows dynamic examination to elicit an unstable apophyseal avulsion.

One particular injury of note is the patellar sleeve fracture. This involves avulsion of the largely unossified distal pole of the patellar in the form of an osteochondral avulsion. Patients present acutely with lipohaemarthrosis and an inability to straight-leg raise. The radiographic findings are those of patellar tendon disruption (high-lying patellar) with an intraarticular fracture (lipohaemarthrosis) and osteochondral fragment (Fig. 2.2). This fragment can be difficult to visualize radiographically although it is well dema

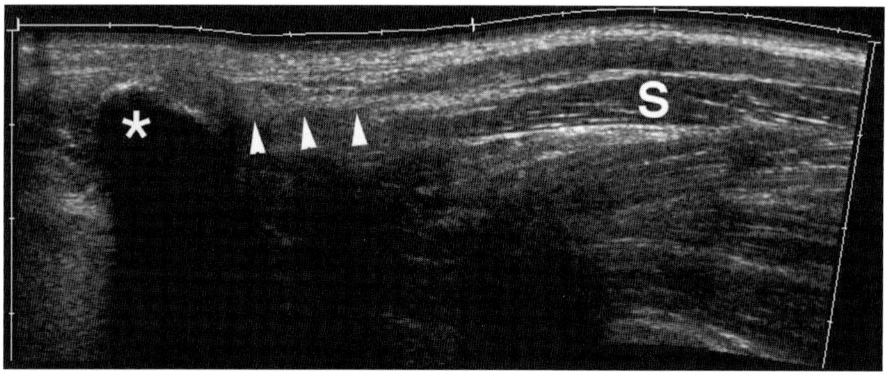

b

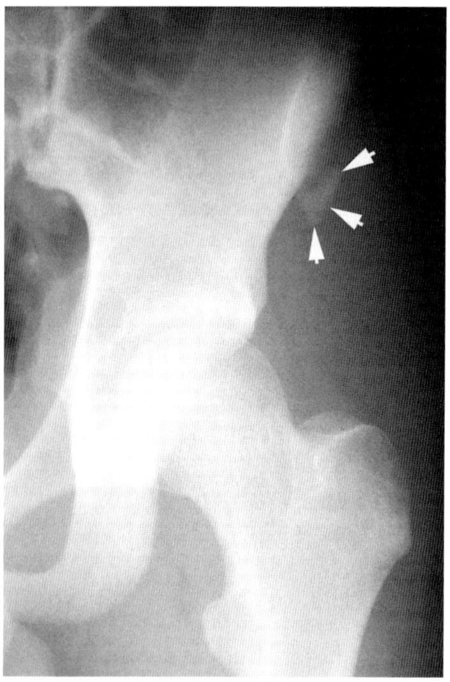

Fig. 2.1a,b. **a** Longitudinal extended field of view sonogram of an acute anterior superior iliac spine (*asterisk*) by the sartorius muscle (*S*). The *arrowheads* outline haematoma tracking deep to the proximal tendon (Image courtesy of Dr P. Robinson, Leeds, UK). **b** Plain radiograph confirming the apophyseal avulsion seen in **a**

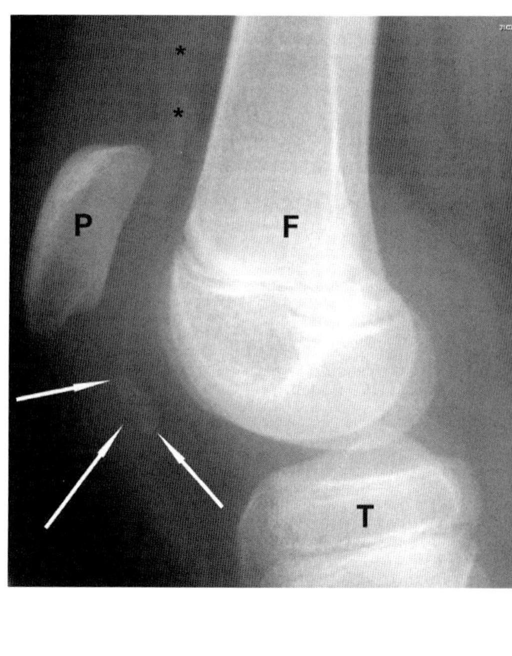

Fig. 2.2 Patellar sleeve fracture in a 9-year-old girl. The patient presented acutely following injury while sprinting with pain and an inability to straight-leg raise. The radiograph shows effusion (*asterisks*) and elevation of the patella (*P*) relative to the femur (*F*). The *arrows* outline faint calcification within an osteochondral fragment

onstrated at US. This injury is differentiated from an osteochondral body by the fact that the patellar tendon remains attached to the fragment (Fig. 2.3).

2.2.1.4 Acute Osteochondral Injuries

Abnormal joint motion leading to shearing, rotatory or impaction forces may fracture one or both of the opposing joint surfaces. Resultant fracture fragments may consist purely of cartilage (chondral fracture) or cartilage attached to a bony fragment (osteochondral fracture). These injuries may be difficult to deduce from the plain radiograph in the acute phase, since a chondral fracture is non-radiopaque, and the density of osteochondral fragments depends on the proportion of subchondral bone they contain. Fracture fragments may become more apparent as loose bodies when their cartilage undergoes degenerative calcification. Osteochondral fragments have characteristic appearances at US with hypoechoic cartilage attached to a variable amount of subchondral bone (Fig. 2.4). Diagnosis of osteochondral fracture therefore requires a high index of clinical suspicion based on knowledge of the mechanism of injury. MR imaging is of value in demonstrating the donor site of displaced fragment. MR imaging can also establish the extent of any bone, cartilage and internal joint disruption, and is useful for planning treatment

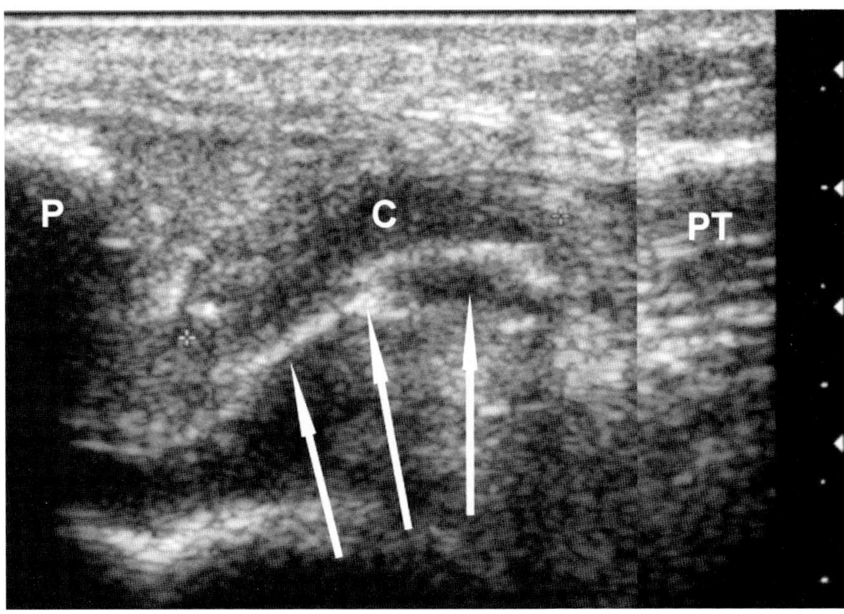

Fig. 2.3 Composite image of the US examination of the patient described in Fig. 2.2. US demonstrates a large cartilaginous component (*C*) compared to the subchondral bone avulsion (*arrows*) with the diagnosis of patellar sleeve fracture confirmed by demonstrating attachment of the patellar tendon (*PT*) to the osteochondral fragment

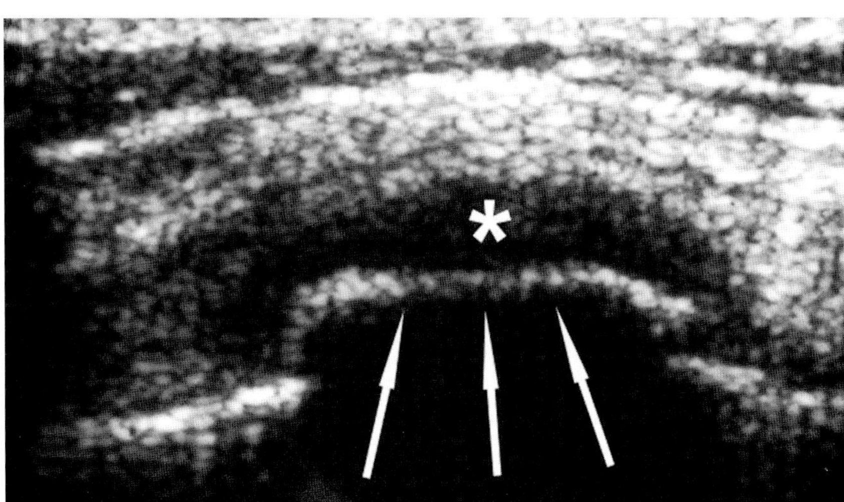

Fig. 2.4 Typical appearance of a displaced osteochondral fragment with avulsed subchondral bone (*arrows*) with a more sizeable cartilaginous component (*asterisk*)

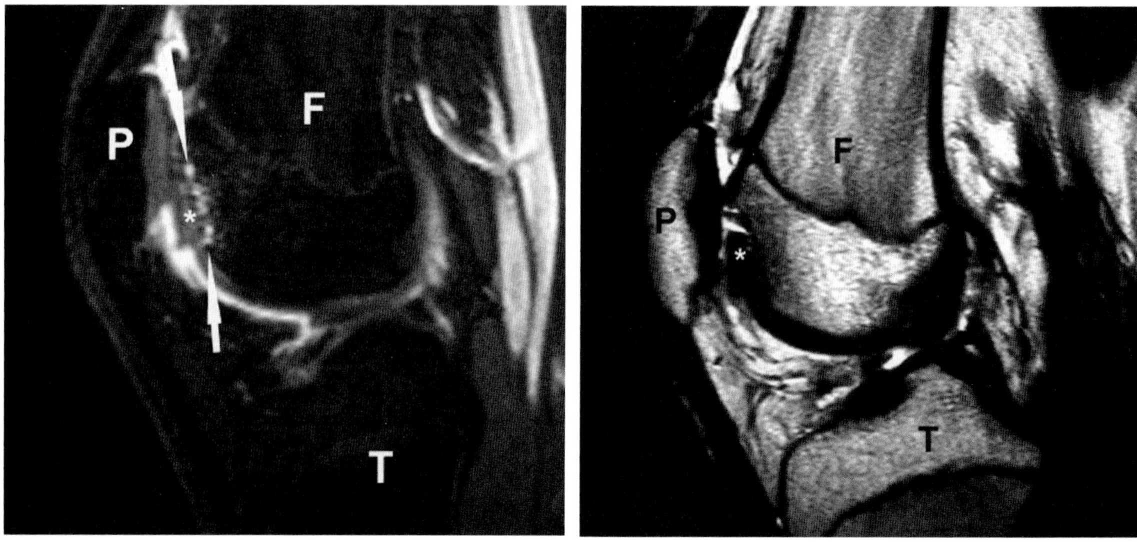

a b

Fig. 2.5a,b. **a** T2-weighted sagittal MRI scan of the knee showing fluid tracking deep to an osteochondral fragment (*asterisk*); *P* patella, *F* femur, *T* tibia. **b** Fat-suppressed T2 sequence showing differentiation between the cartilaginous and bone components of the fragment (*asterisk*) with a fluid track beneath the partially displaced fracture (*arrows*); *P* patella, *F* femur, *T* tibia

(Fig. 2.5). Most non-displaced lesions in patients with open physis will heal with conservative management, but displaced fragments or skeletal maturity often require surgical intervention [10].

US is becoming increasingly useful in the emergency setting. It is well tolerated by an injured child who may find positioning for radiographs distressing. Certainly, it is preferable to MR imaging which may require a general anaesthetic. Emergency physicians have found it to be helpful in guiding the reduction of difficult paediatric forearm fractures [11]. One emergency department has described the use of US to demonstrate uncomplicated greenstick and torus fractures [12]. There has been a case report of the detection of a radiographically occult fracture of the radial neck with US [13]. A large study of 224 suspected fractures in children found that US was most reliable for detection of simple femoral and humeral diaphyseal fractures and fractures of the forearm. It was less dependable for compound injuries, non-displaced epiphyseal fractures (Salter-Harris type 1) and those with a fracture line of less than 1 mm [14].

A particular area of diagnostic difficulty is the elbow in children. Supracondylar fractures carry a high rate of complications including neurovascular compromise. There may be few clues to the presence of a supracondylar fracture on the plain film, particularly when the fracture involves the cartilaginous growth plate. Superior displacement of the coronoid fat pad is the most sensitive indicator of a traumatic joint effusion. However, this may be missed if an adequate lateral radiograph is not obtained. Transphyseal supracondylar fractures can be clearly shown with US [15]. Unstable lateral humeral condyle fracture lines involve the cartilaginous trochlea and cannot be identified on plain films. MRI has been considered necessary to distinguish between the conservatively treated stable fracture, and the surgically treated unstable fracture, which extends to the joint surface. However, ultrasonography has recently been shown to be capable of ruling out joint involvement [16]. The diagnosis of fracture in a child's elbow is further complicated by the variable ages at which the ossification centres are visible on plain films. Inexperienced casualty officers may not be aware of the order of epiphyseal ossification around the elbow (Table 2.2), and a displaced medial humeral epiphysis may be easily overlooked with catastrophic results. Ultrasonography can be used to study the

Table 2.2 Order and approximate timings of secondary ossification site around the elbow

Capitellum	1 year (3–6 months)
Radial head	4 years (2–6 years)
Medial epicondyle[a]	7 years (4–6 years)
Trochlea	10 years (9–10 years)
Olecranon	10 years (6–10 years)
Lateral epicondyle	11 years (9–12 years)

[a]Last to fuse

hypoechoic, unossified epiphysis around the elbow and to demonstrate physeal separations [17, 18].

2.2.2 Foreign Bodies

Foreign body demonstration is notoriously difficult with radiographs particularly with wooden splinters, which are non-radiopaque, and can cause significant morbidity if they remain undetected. Wood, plastic and glass splinters are usually highly echogenic when examined with US. It is possible to localize foreign bodies as small as 1 to 2 mm using a high frequency linear transducer [19]. The use of US to detect and remove foreign bodies by emergency physicians has been described [20] and intraoperative US by surgeons has been recommended by LEUNG et al. [21] to ensure complete removal.

2.2.3 Haematoma

Haematomas are common after a myotendinous injury or a direct blow and are usually intramuscular in location. MR findings may vary depending on the time elapsed since the injury, but usually show characteristics of methaemoglobin with increased signal on both T1- and T2-weighted sequences [22]. Haematomas tend to resorb over a period of 6 to 8 weeks, but may occasionally leave serous fluid within a connective tissue sheath. These pseudotumours are seen mainly in the rectus femoris but also in semimembranosus and semitendinosus muscles. US findings in haematoma are dependent on the time of imaging following injury. In the acute phase, haematoma may be echogenic and difficult to discriminate from normal muscle (Fig. 2.6a) [19]. Later, US may show a complex ovoid mass with internal septations due to clot

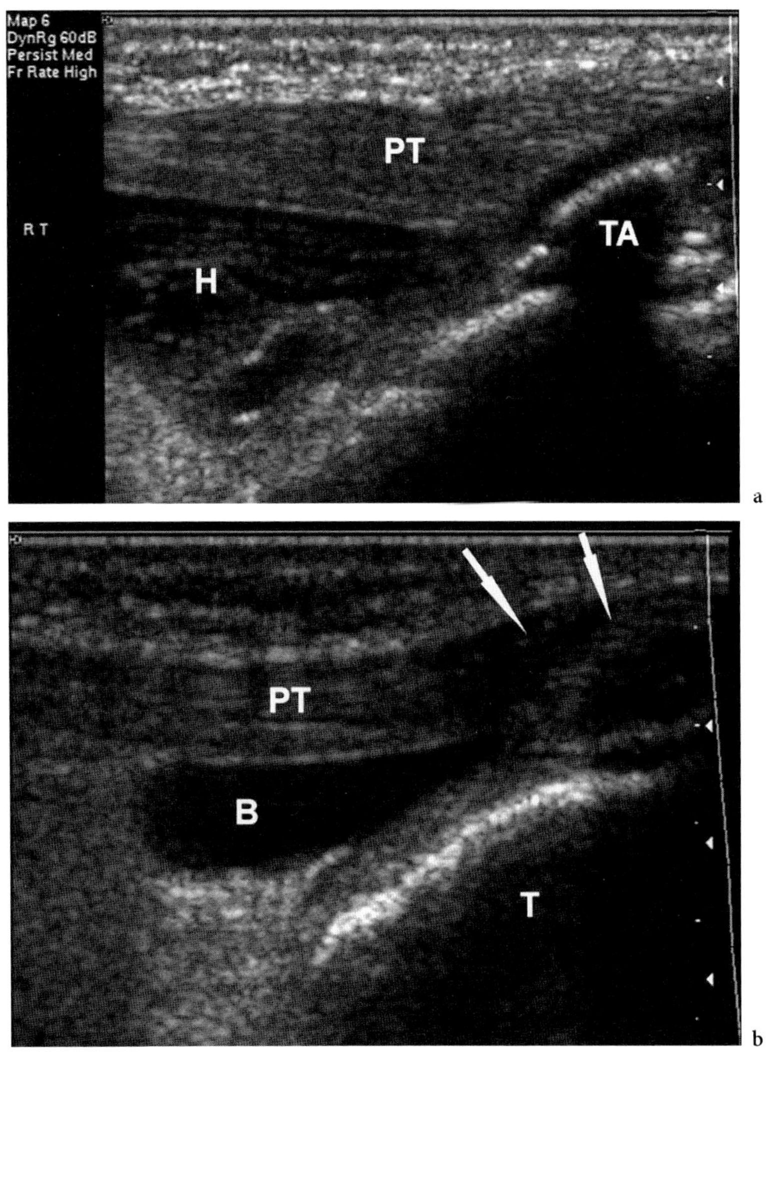

Fig. 2.6a,b. a Acute Osgood-Schlatter disease with acute avulsion of the tibial tuberosity apophysis (*TA*) during a sprinting race. US on the day of injury shows haematoma (*H*) in the deep infrapatellar bursa (*PT*). The acute haematoma (*H*) has the classical appearance of mixed hyper and hypoechogenicity. **b** 10 days after injury the haematoma seen within the deep infrapatellar bursa in **b** has an altered appearance having become anechoic (*B*) in keeping with resolving haematoma. Thickening of the distal patellar tendon (*PT*) is now present (*arrows*); *T* tibia

retraction eventually becoming anechoic (Fig. 2.6b). There is usually a lack of inflammatory change in the surrounding subcutaneous soft tissues.

2.2.4 Muscles and Tendons

Muscle and tendon injuries are uncommon in the immature skeleton since the weakest link in the muscle-tendon-bone chain is the growth plate. However, young athletes may incur sprains and tears, particularly at myotendinous junctions, although complete tendon rupture is rare. Strains most often affect muscles that cross two joints since these are susceptible to greater degrees of stretch [6]. The three hamstring muscles, particularly the biceps femoris, are the most commonly injured muscles in jumpers and sprinters [22]. Each hamstring myotendinous junction extends almost the full length of the muscle so that when injury occurs it can be located at either end of or within the muscle belly. US allows dynamic assessment of tendon and muscle injury. Tendon injury is unusual in this age group and will not be discussed in depth.

The dynamic element of US is vital in the functional assessment of tendon injury helping differentiate partial from full-thickness injury (Fig. 2.7). The same is true of myotendinous junction injury (Fig. 2.8) with active contraction giving an excellent assessment of the degree of disruption present (Fig. 2.9). It should be remembered that US demonstrates well any structural disruption of the myotendinous junction but does not demonstrate muscle oedema to great effect. This in part results from muscle anisotropy making the assessment of muscle echotexture difficult. This means that US has a low sensitivity for grade 1 tears and really only demonstrates the size of the haematoma and the degree of structural disruption. MRI, however, has difficulty separating muscle oedema, haematoma and structural disruption, and as a result tends to overestimate the severity of the injury (Fig. 2.10). These methods must be used in combination to achieve an accurate assessment of muscle injury.

2.2.4.1 Shoulder

Young athletes involved in throwing, swimming and racquet sports are at risk of shoulder impingement syndrome and rotator cuff tendinopathy [23]. Tendinopathy is infrequently seen with the major-

a

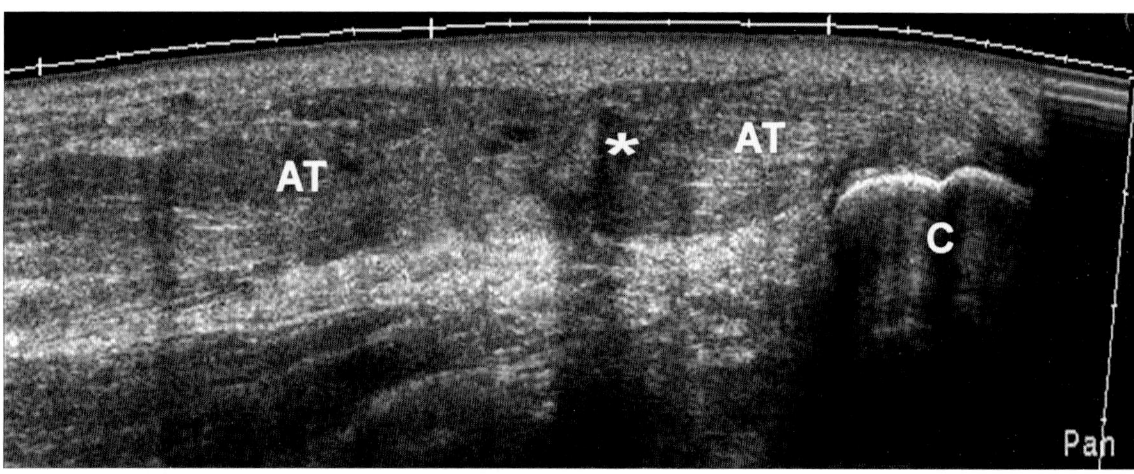

b

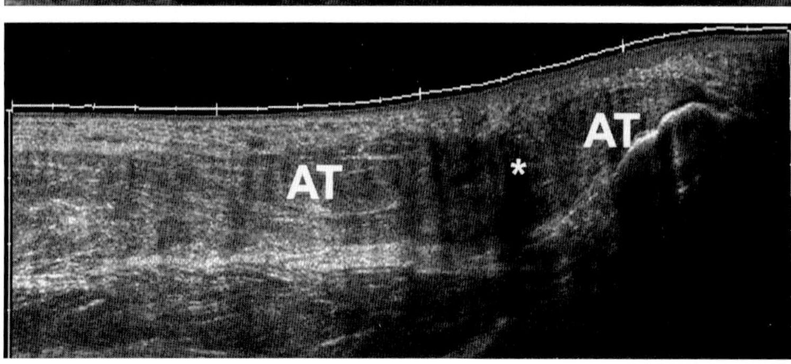

Fig. 2.7a,b Complete rupture of the Achilles tendon (*AT*) with differential movement seen between the proximal and distal portions during movement. The tear (*asterisk*) is best demonstrated during plantar flexion; *C* calcaneus

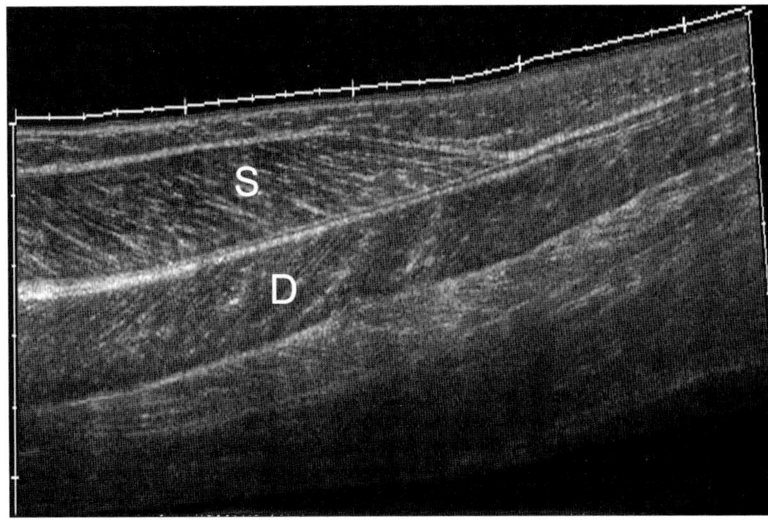

Fig. 2.8 Extended field of view sonogram showing the normal myotendinous junction. Superficial (*S*) and deep (*D*) components feed on the aponeurosis which gradually thickens distally to form a tendon

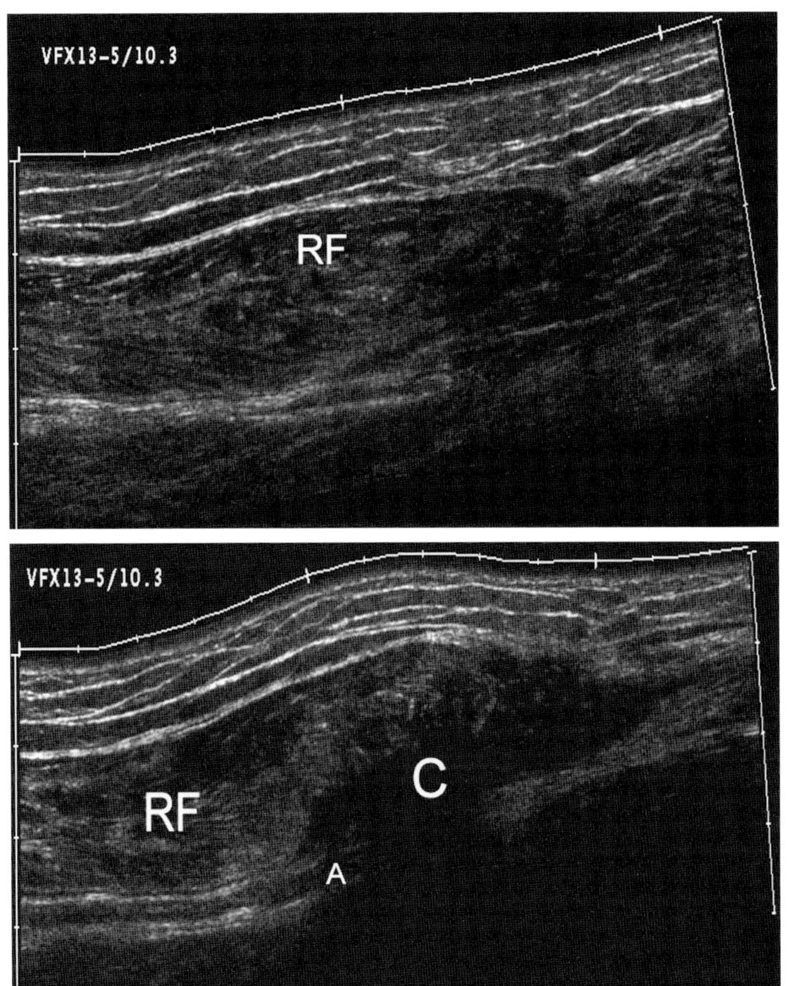

Fig. 2.9a,b Extended field of view sonograms of the thigh at rest (**a**) and during quadriceps contraction (**b**) showing improved demonstration of the extent of myotendinous disruption (*C*) with contraction. The rectus femoris muscle (*RF*) is seen to pull away from the underlying aponeurosis (*A*)

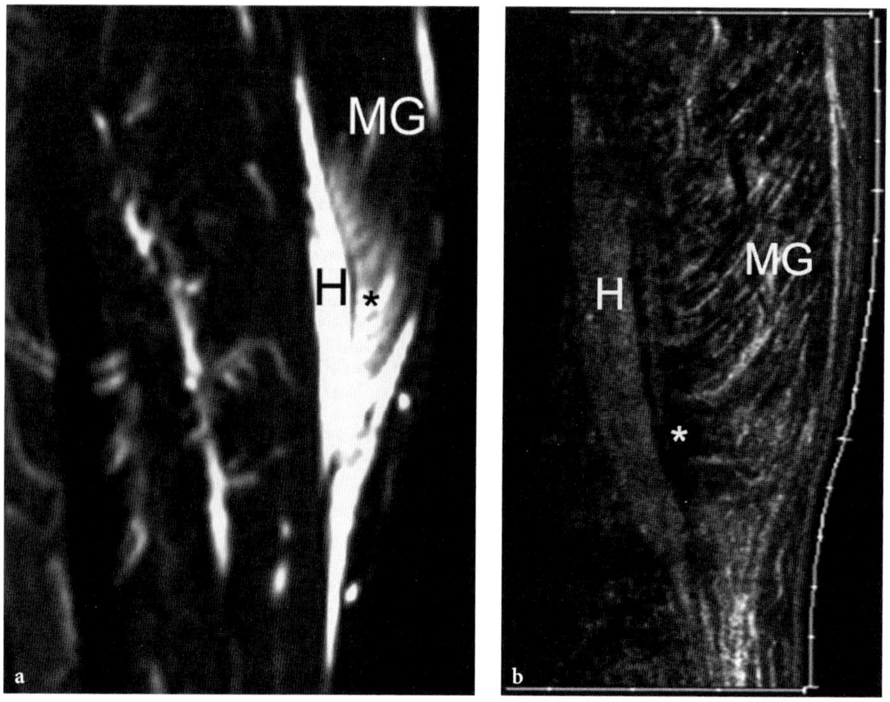

Fig. 2.10a,b Coronal T2 fat-suppressed MR and US images of the distal medial head of gastrocnemius (*MG*) performed on the day of injury. Both demonstrate the degree of myotendinous disruption and haematoma formation (*H*) with oedema within the adjacent muscle best demonstrated with MR (*asterisk*). MR however has difficulty differentiating haematoma and fluid from oedema and although more sensitive than US can overestimate the degree of injury

ity of patients presenting with bursitis. This is most commonly seen with overhead sporting activities such as tennis, swimming (crawl, butterfly and back stroke) and throwing, and is associated with hypermobility and relative dysfunction of the static stabilizers of the shoulder. Both MR and US can be used to assess the rotator cuff tendons, but US has the advantage that it allows dynamic assessment of the shoulder. The typical appearance is that of fluid trapping in the subacromial bursa either side if the coracoacromial ligament (Fig. 2.11). US can also be used to guide needle placement of steroid into the subacromial bursa during treatment of impingement. This must be performed with caution in this age group and steroid should used like a long-acting local anaesthetic to permit rehabilitation. Care should be taken with regard to the doping regulations of the specific sport as permission to use steroid and formal notification may be required. Rest, training modification and physiotherapy are the mainstays of treatment.

2.3 Chronic Trauma

Overuse injuries are common in young athletes. They can result in growth arrest, osteochondroses, or stress fractures, and are therefore potentially serious diseases with long-term sequelae.

2.3.1 Chronic Fracture Patterns

Young gymnasts who repeatedly stress their wrists with work on the bars and beam are subject to chronic Salter-Harris type 1 stress fractures through the wrist growth plates. These lesions resemble rickets on plain radiographs but they do not occur in the other joints and the serum biochemistry is normal [7]. A variant of this wrist injury is seen in skateboarders in whom repeated falls preferentially onto one wrist produce unilateral Salter-Harris type 1 fractures (Fig. 2.12a,b). US is rarely of value in the diagnosis of Salter-Harris injury but can be of value where there are borderline plain film abnormalities present.

Slipped capital femoral epiphysis is the most important chronic Salter-Harris injury seen. These patients can be difficult to diagnose and present with either hip, knee or thigh pain. It is not unusual for US to be a first-line investigation for hip pain in young patients. As such, a high index of suspicion must be maintained for this condition when performing hip examinations in the skeletally immature (Fig. 2.13).

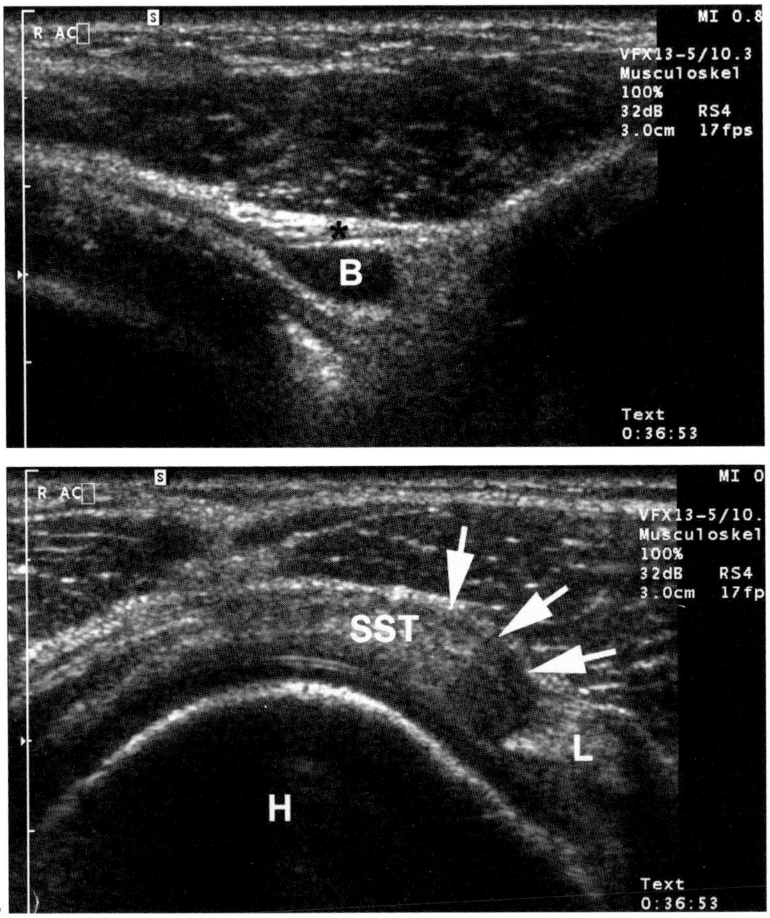

Fig. 2.11a,b. Competitive 11-year-old swimmer with shoulder pain. Sonograms through the coracoacromial ligament (*B*) demonstrate the ligament (*asterisk*) with fluid pooling deep to the ligament (*B*) in keeping with impingement. Transverse sonograms of the supraspinatus (*SST*) confirm this bursal thickening present over the anterior free edge of the SST (*arrows*) adjacent to the rotator cuff interval; *H* humeral head, *L* long head of biceps tendon

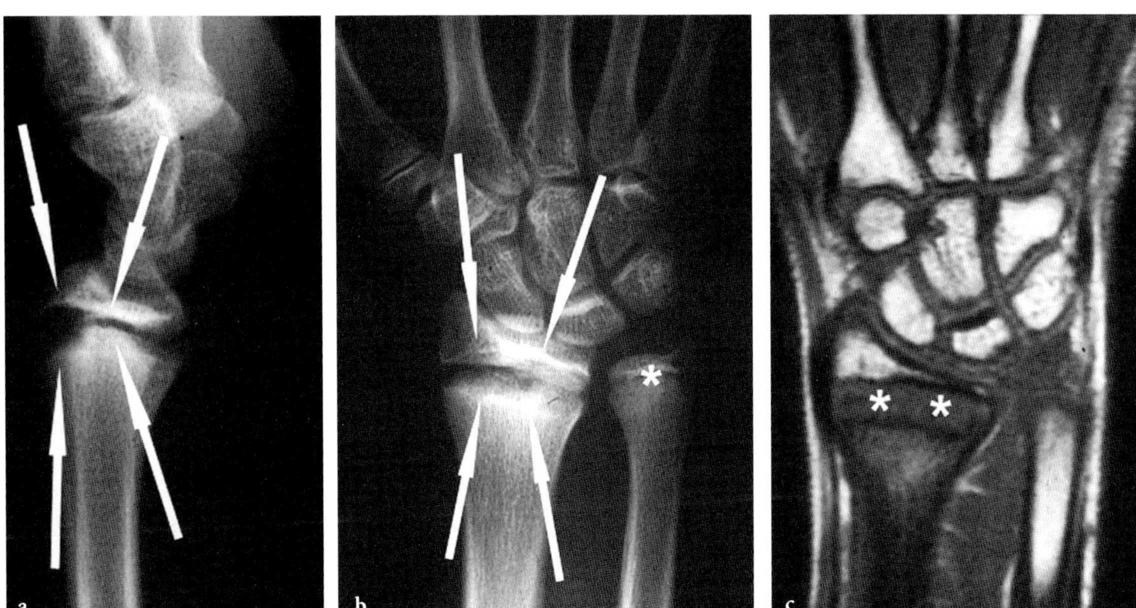

Fig. 2.12a–c. Chronic Salter-Harris 1 fracture of the wrist in a gymnast. The classical widening of the growth plate (*arrows*) with adjacent sclerosis of the distal radius. Similar though less-marked changes are seen in **b** in the distal ulna (*asterisk*). T1-weighted coronal MR (c) shows widening of the growth plate (*asterisks*) with sclerotic margins

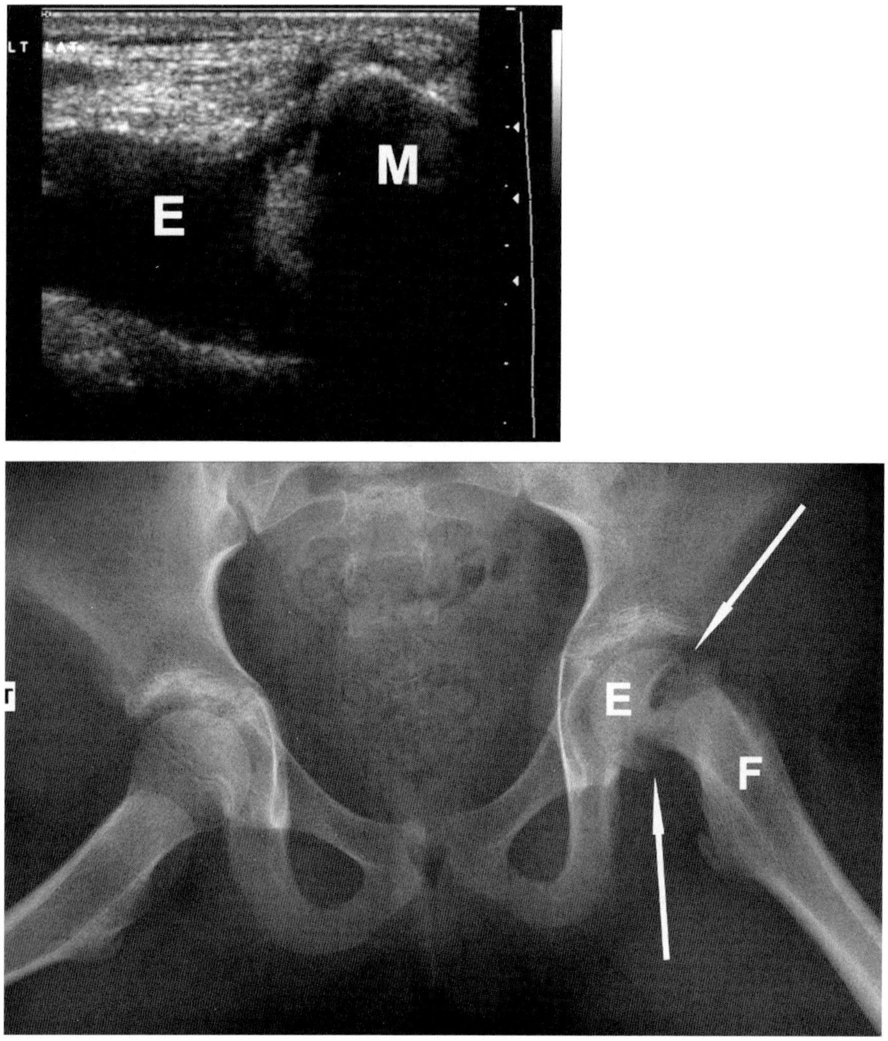

Fig. 2.13a,b. Adolescent patient presenting with chronic hip pain. The initial US (a) demonstrated an effusion (*E*) with a bare metaphysis (*M*). Subsequent radiographs (b) demonstrated the slipped capital femoral epiphysis (*arrows*); *F* femoral metaphysis, *E* proximal femoral epiphysis

Chronic avulsion injuries resulting from repetitive trauma may heal with bubbly sclerosis resembling tumour, particularly around the ischium. The histology of these lesions can be confusing, and biopsy should be avoided. Clinical history will normally point towards chronic repetitive trauma, and MR imaging is helpful in clarifying the diagnosis [8].

Overuse injuries to the bone can result in stress fractures when repetitive mechanical loading overcomes the body's healing capacity. One-third of stress fractures occur in athletes between 16 and 19 years of age. They are more common in female athletes [24], and most are found in the lower limb, particularly the proximal tibia [25]. Typically, the pain is relieved by rest and reoccurs with the resumption of activity, although rest pain may be present in severe injuries [26]. Diagnosis requires a high index of suspicion since the initial radiographs are usually normal. Films taken 2 to 3 weeks after the onset of symptoms may show localized periosteal and endosteal thickening. The differential at diagnosis at this stage is wide and includes osteoid osteoma, Ewing's sarcoma and eosinophilic granuloma. Sequential radiographs and a careful clinical history normally help differentiate stress fractures from other lesions. Progression to a full-thickness fracture can occur if the fracture is not allowed to heal with rest. US can demonstrate periosteal reaction and new bone formation in stress injury to bone although it is not a primary diagnostic tool. Plain films are the first investigation which if negative should be followed up with either MRI or scintigraphy.

Bone scintigraphy will show increased uptake 2 to 3 days after the onset of symptoms (Fig. 2.14). The earliest changes of a stress reaction are perios-

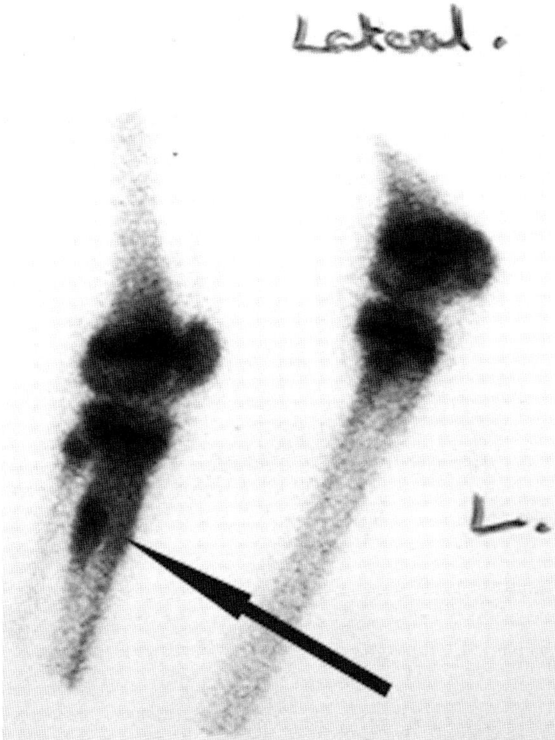

Fig. 2.14 Bone scintigraphy demonstrating focal increased activity (*arrow*) in a stress fracture on the proximal tibia

teal, muscle and bone marrow oedema, and these are clearly demonstrated with MR using a STIR (short tau inversion recovery) or fat-suppressed T2 sequence (Fig. 2.15) [26]. Later, frank stress fractures are seen as linear bands of low signal on T1-weighted sequences (Fig. 2.16) [7]. It should be remembered that scintigraphy is non-specific and will show increased activity related to periosteal traction injury and the site of tendon and muscle insertions as well as in stress injury to the underlying bone. Scintigraphy's lack of specificity means that in general MRI is of greater diagnostic value as it differentiates between periosteal, cortical and medullary disruption (Fig. 2.15).

CT is not generally employed to diagnose stress fractures of the appendicular skeleton since it is less sensitive than bone scintigraphy and MR. Occasionally, stress fractures in children may exhibit marked periosteal proliferation mimicking tumour. The CT demonstration of endosteal bone formation in these cases confirms the presence of a stress injury [27]. CT is particularly useful when stress fractures of the pars interarticularis are suspected. Spondylolysis occurs in the lumbar spine as a result of repeated hyperextension, particularly in cricketers, and presents with back pain around the time of the adolescent growth spurt. Plain radiographs may show the fracture through the pars interarticularis. However, local sclerosis may suggest the possibility of osteoid osteoma. Bone scintigraphy is of little help as a discriminator since uptake is increased in both conditions. At this point, thin-section "reverse-angle" CT will demonstrate the bony nidus of osteoid osteoma or the fracture line of spondylolysis.

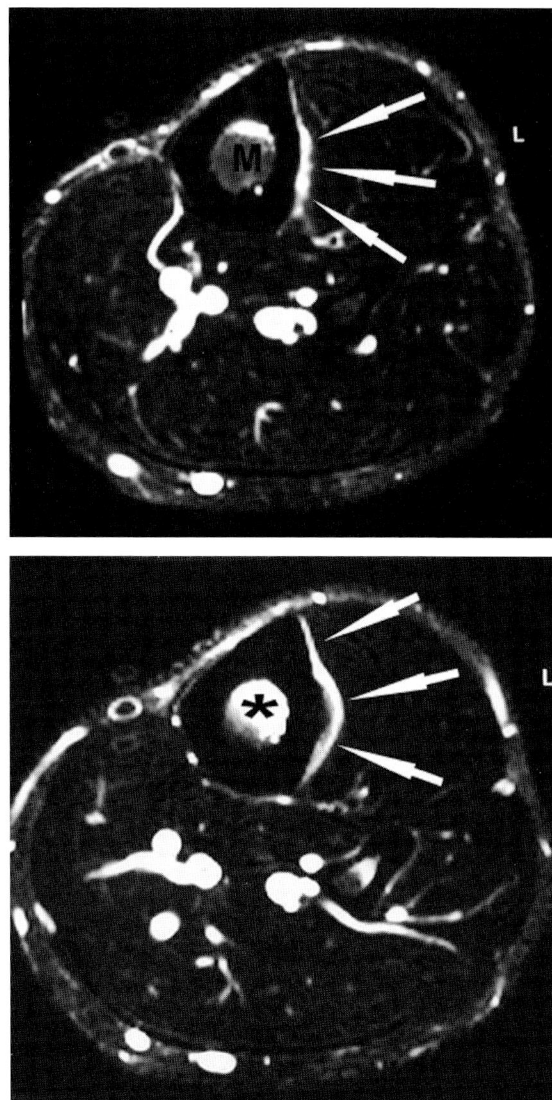

a

b

Fig. 2.15a,b. T2 fat-suppressed axial MR images through the tibia demonstrating differing degrees of stress injury to the tibia. **a** Mainly periosteal oedema (*arrows*) with little cortical thickening and minor medullary oedema (*M*). **b** More advanced change with periosteal oedema (*arrows*), cortical thickening and marked oedema of the medullary cavity (*asterisk*) in keeping with a developing stress fracture

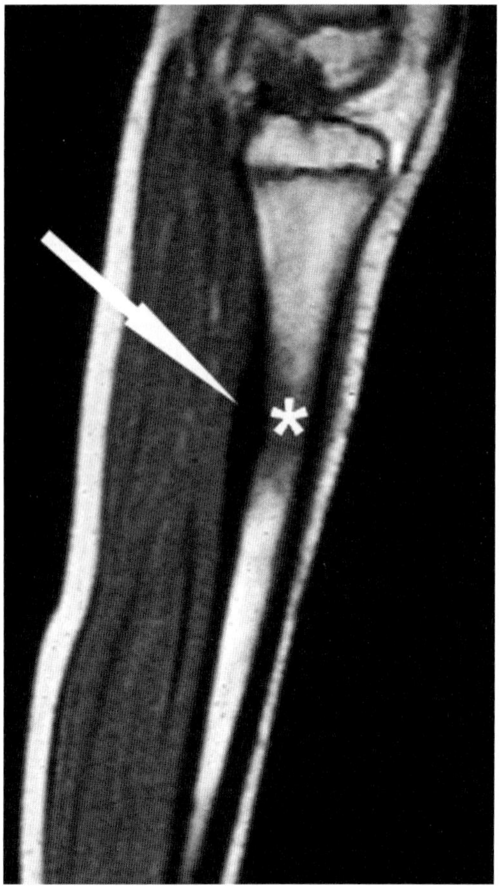

Fig. 2.16 T1-weighted sagittal MR scan of the tibia demonstrating an established stress fracture (*asterisk*) with marked periosteal new bone formation (*arrow*)

2.4 The Osteochondroses

There are a number of osteochondroses reported in children and adolescents, many with eponymous names, and confusing descriptions of disease and aetiology in the literature. Kohler's disease results in osteonecrosis of the tarsal navicular while the surrounding cartilage is preserved. Freiberg's disease is osteonecrosis of the metatarsal head following osteochondral fracture. Sever's disease is described as an apophysitis of the posterior calcaneal apophysis [28]. The common factor appears to be repetitive trauma, and the diagnosis can usually be made from the plain radiograph. Most paediatric osteochondroses heal with conservative treatment, and little long-term morbidity.

2.4.1 Osteochondritis Dissecans

Osteochondritis dissecans (OCD) is a lesion of unknown cause, which results in an island of abnormal subchondral bone separating from the normal bone. It may also break away entirely, leading to an osteochondral loose body which can interfere with joint function. While the exact aetiology is not clear, the common denominator appears to be overuse [23], and it is mainly seen in adolescent athletes. The common sites are the capitellum, the talus and the medial femoral condyle. OCD is usually apparent on the plain radiograph, but MR imaging will show subtle bone oedema at an early stage [8]. Management of these lesions partly depends on knowing whether the articular cartilage overlying the separated fragment of subchondral bone is intact [7]. MRI showing high-signal oedema on T2-weighted sequences between the fragment and the underlying bone indicates fragment instability [29]. Takahara et al. [30] have demonstrated that sonography can depict unstable osteochondral lesions in the elbow as discrete echogenic intra-articular foci. If the fragment can be shown to be loose, then surgical fixation may be required to reduce the risk of impaired joint function (Fig. 2.17).

Sinding-Larsen disease is thought to arise from repetitive traction by the patella tendon leading to irregular calcification and ossification of the inferior pole of the patella (Fig. 2.18). It is commonest between the ages of 10 and 12 years. Osgood-Schlatter disease effects preteen and early teenage athletes with a slight preponderance in boys, and seems to be related to repetitive squatting or jumping [5]. It is thought to be due to chronic traction by the inferior patella tendon on the tibial tuberosity. The diagnosis is usually clinical, but radiographs and MRI can be used to exclude other causes of knee pain. US may show thickening of the patella tendon, which may appear indistinct, and partly echogenic. Fragmentation of the tibial tuberosity and hypoechoic surrounding soft tissue oedema may also be present [31]. Osgood-Schlatter disease can present either as an acute avulsion with surrounding haematoma (see Fig. 2.6a) or, as is more common, as a chronic pain syndrome (see Fig. 2.6b). In the chronic setting pain can result either from traction and fragmentation of the tuberosity or from local soft-tissue change. This is usually related either to deep infrapatellar bursitis of from pressure effects causing formation of a superficial adventitial bursa.

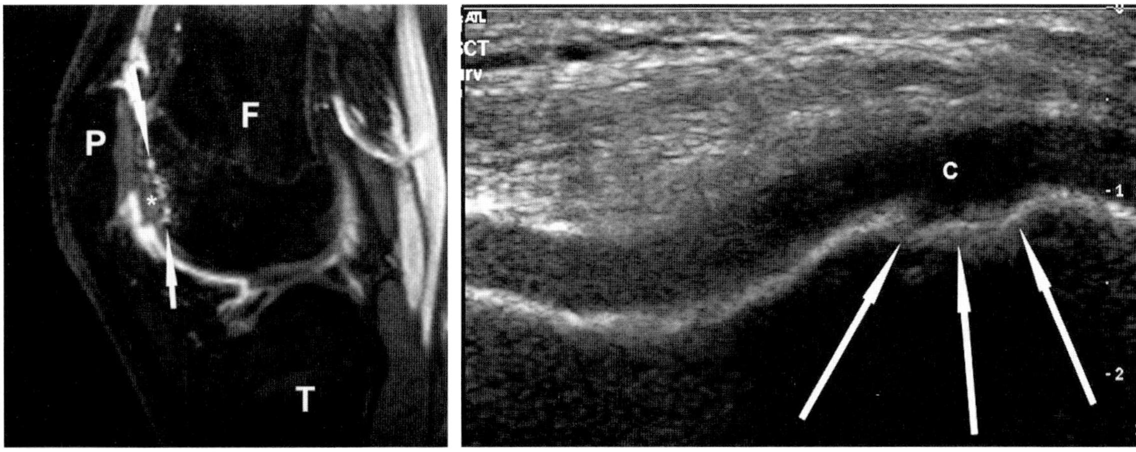

a b

Fig. 2.17a,b. a T1-weighted coronal MR of the knee showing an osteochondral lesion of the medial femoral condyle (*asterisk*); *F* femur, *T* tibia. **b** US examination performed in extreme flexion allows demonstration of the osteochondral lesion seen in **a**. This demonstrates the subchondral collapse (*arrows*) and shows intact overlying cartilage (*C*) (Image courtesy of Dr. A.J. Grainger, Leeds, UK)

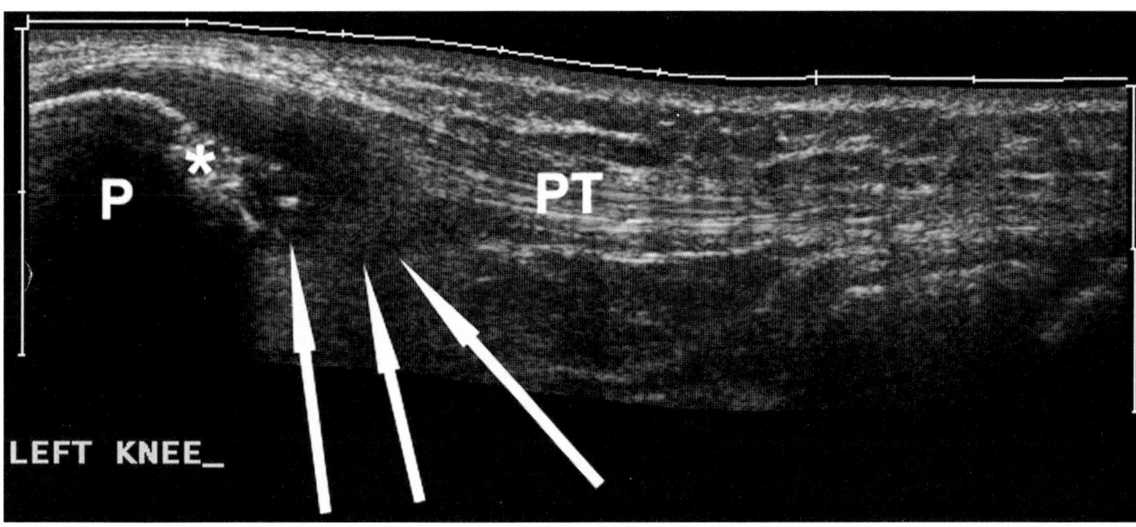

Fig. 2.18 Extended field of view sonogram of the proximal patellar tendon in a keen skate-boarder showing thickening (*arrows*) of the proximal patellar tendon (*PT*) with fragmentation and erosion (*asterisk*) of the lower pole of the patella (*P*) in keeping with Sinding-Larsen or jumper's knee

2.4.2
Panner's Disease

Panner's disease is a developmental osteochondrosis of the capitellum seen in children between 6 and 12 years of age and almost exclusively in baseball pitchers [23]. The repeated throwing action is thought to set up compressive strains in the radiocapitellar joint with disruption of the medial epicondyle growth plate or the ulnar collateral ligament. The repeated compression of the radiocapitellar joint is thought to compromise the blood supply of the subchondral plate causing avascular necrosis of the capitellum. The appearance of the plain film is said to be diagnostic with sclerosis and flattening of the capitellum, and roughening of its articular surface. MRI may demonstrate fragmentation and decreased signal on T1-weighted sequences of the ossifying capitellum [32]. Loose bodies are unusual in Panner's disease [33]. Unlike osteochondritis dissecans, it resolves with rest, and involves a younger age group.

2.4.3 Medial Epicondylitis (Little League Elbow)

Baseball pitchers between 8 and 12 years of age experience repeated valgus stress to the medial elbow during the throwing motion, causing breakdown of the physis attaching the medial epicondyle to the humerus. This has an association with capitellar osteochondral injury, and the nomenclature regarding these two conditions can be confusing. Regardless of what one calls them, they are essentially the same process. This is a true apophysitis, as distinct from the adult golfer's elbow which results from flexor-pronator tendonitis [23]. MR may show marrow oedema and irregularity of the physis, whilst skeletal scintigraphy demonstrates asymmetrical increased uptake in the symptomatic medial epicondyle [33].

2.5 Accessory Ossicles

Accessory ossicles are mostly described as normal variants, although there is no doubt that some may become symptomatic. They are joined to normal bone by fibrous tissue, which can lead to the development of a painful pseudarthrosis if disturbed by frequent, vigorous exercise [7]. The os trigonum posterior to the talus is a commonly reported source of pain, particularly in young gymnasts and dancers (Fig. 2.19). This is thought to be secondary to repetitive impaction of the os trigonum between the calcaneus and posterior tibial malleolus during plantar flexion [33]. The os trigonum may even develop as a result of impingement with the posterior portion of the talus more prone to damage and fragmentation as it is the last portion of the talus to ossify (Fig. 2.20). Radiographs are usually unhelpful, but bone scintigraphy may demonstrate increased uptake compared to the asymptomatic side. MR will show bone oedema (Fig. 2.21) [34].

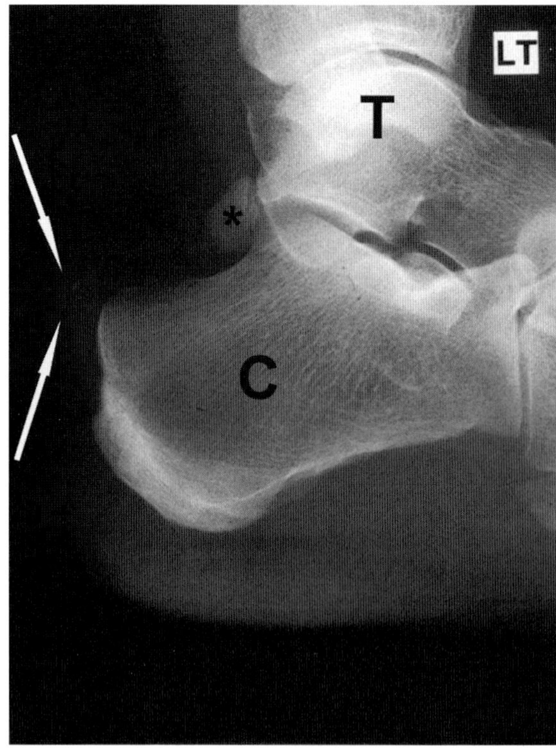

Fig. 2.19 Lateral ankle radiograph showing an os trigonum (*asterisk*) with some incidental calcification noted in the distal Achilles tendon (*arrows*); *C* calcaneus, *T* talus

Other common symptomatic accessory ossicles seen in adolescent patients are the os tibiale externum at the site of the tibialis posterior insertion on the navicular and the bipartite patellar. These syndesmoses can become disrupted and symptomatic. Using US the site of the syndesmosis can be located and the patient's symptoms correlated. Final confirmation of movement producing bone oedema around the syndesmosis can be achieved using either scintigraphy or fat-suppressed MRI. US can be used to perform guided injections of the syndesmosis with steroid and local anaesthetic either for diagnostic or therapeutic purposes (Fig. 2.22).

2.6 Bursae

Bursal inflammation can be seen in the adolescent athlete, although it is normally associated with an underlying bone or biomechanical abnormality. Osteocartilaginous exostoses are the commonest cause of local irritation. They cause compression and displacement of adjacent structures which, in association with activity, results in bursa formation and local pain. The commonest site for this is around the knee where femoral osteocartilaginous exostoses cause irritation to the overlying quadriceps (Fig. 2.23) and around the medial tibial metaphysic where exostoses can produce pes anserinus irritation (Fig. 2.24)

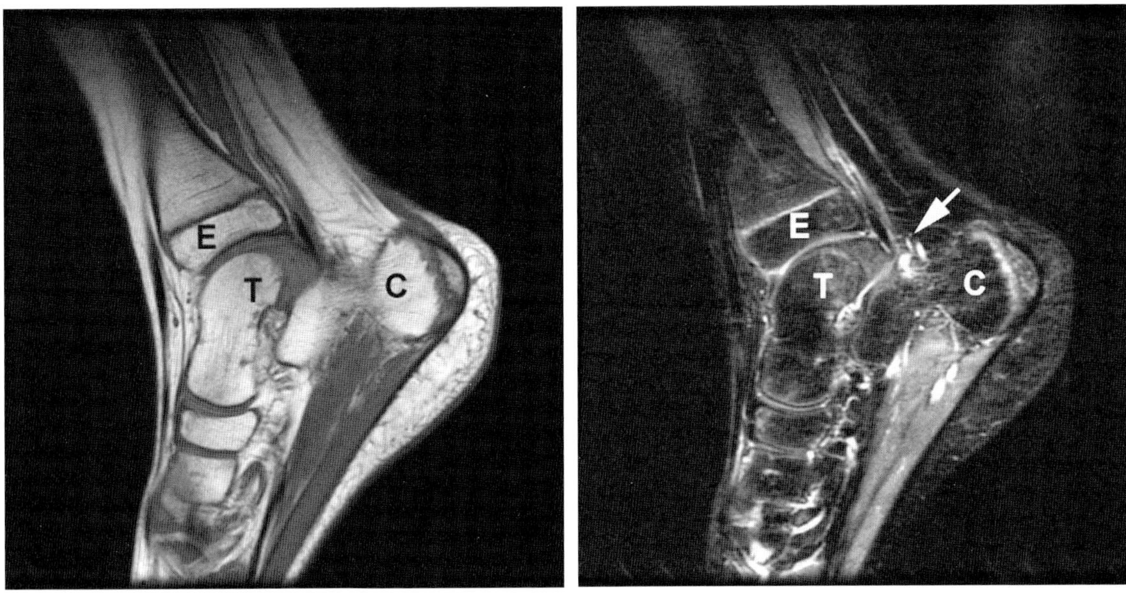

a b

Fig. 2.20a,b. T1 and T2 fat-suppressed sagittal MR scans of the ankle in an adolescent. These demonstrate that the posterior talus is the final portion to ossify; *E* distal tibial epiphysis, *T* talus, *C* calcaneus

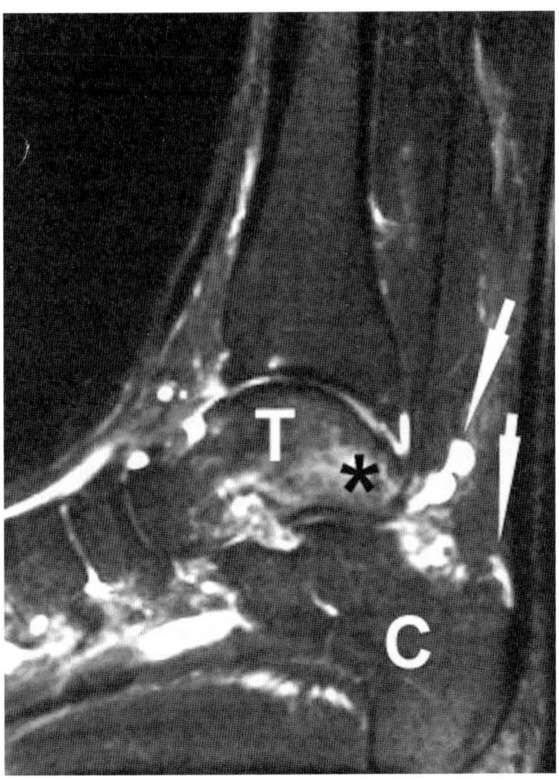

Fig. 2.21 T2-weighted sagittal MR scan of the ankle. There is symptomatic posterior impingement of the ankle with oedema present in the posterior talus (*asterisk*). Fluid is also present in the posterior recess and retrocalcaneal bursa (*arrows*); *T* talus, *C* calcaneus

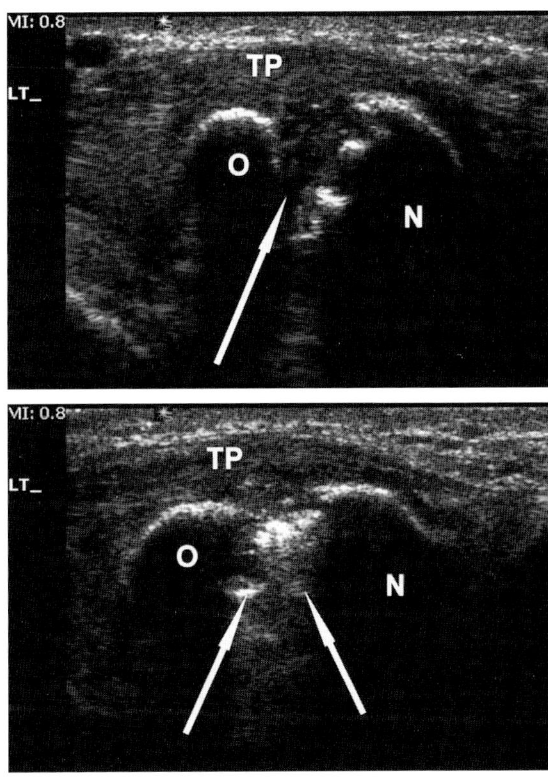

a b

Fig. 2.22a,b. Symptomatic os tibiale externum in a footballer. **a** The synchondrosis (*arrow*) is well demonstrated between the navicular (*N*) and the ossicle (*O*) with the tibialis posterior tendon (*TP*) lying superficially. **b** After guided injection the synchondrosis has become filled with echogenic steroid (*arrows*)

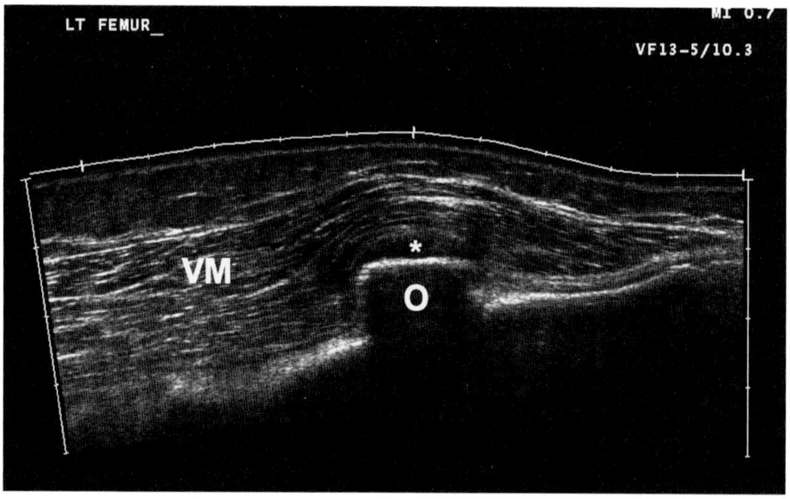

Fig. 2.23 Extended field of view sonogram of a distal femoral osteocartilaginous exostosis (*O*) indenting the deep aspect of the vastus medialis (*VM*). The cartilage cap of the exostosis is well demonstrated (*asterisk*)

a

b

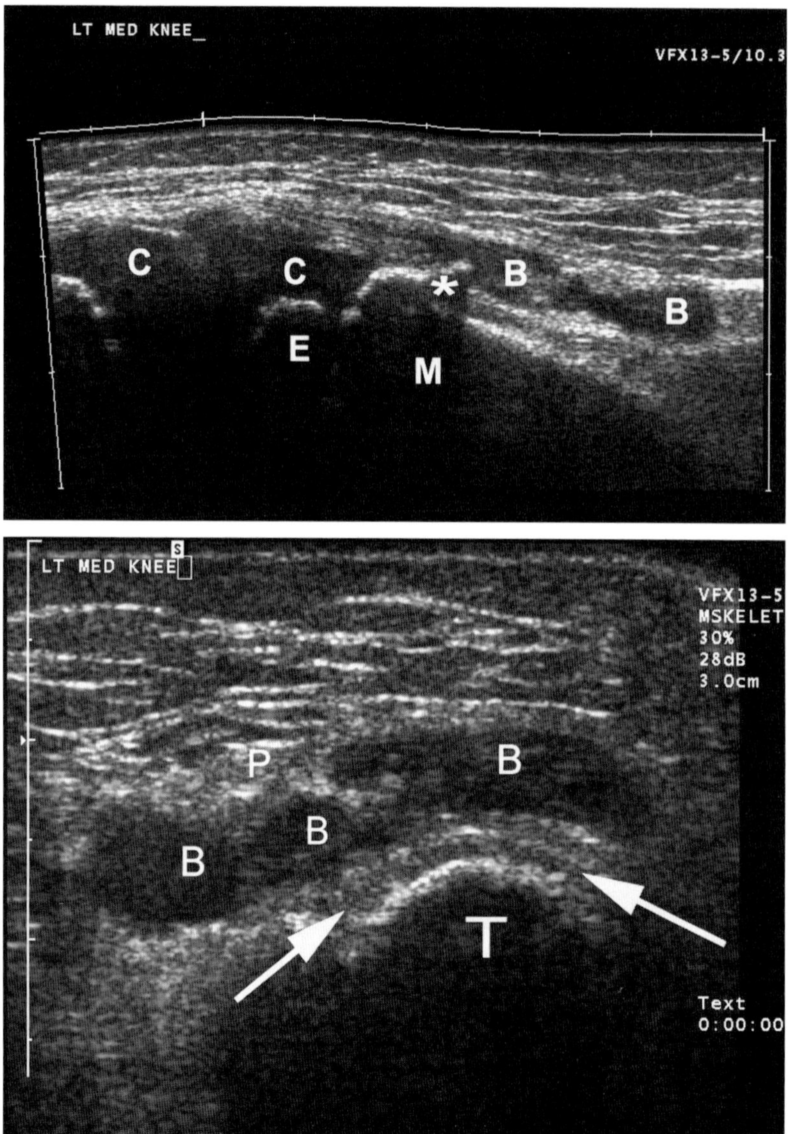

Fig. 2.24a,b Extended field of view sonograms (**a** longitudinal, **b** transverse) of the medial aspect of the knee in an 8-year-old footballer. There is pes anserinus bursitis (*B*) overlying the distal portion of the MCL (*arrows*). The underlying cause for this is a small exostosis (*asterisk*) which is demonstrated on the longitudinal scans indenting the deep aspect of the bursa; *C* cartilage of the femoral and tibial epiphyses, *E* tibial epiphysis, *M* medial tibial metaphysis, *T* tibia

2.7 Summary

There are fundamental differences in injuries sustained during trauma and sport between children and the skeletally mature adult. The suspected type of injury dictates imaging needs, and this requires close communication between the referring clinician and the radiologist. The use of US has expanded rapidly over the last decade with the advent of high-resolution transducers, which allow tissues to be demonstrated in exquisite detail. It is an excellent tool for assessment of the immature skeleton where it can easily distinguish cartilage from bone, and bone from soft tissue. Extended field of view techniques are able to display large continuous sections of anatomy to show the relationship of diseased to normal tissues. It is possible to compare appearance with the asymptomatic side, and to observe the tissues during joint motion. US is well tolerated by the child, and is readily available in most departments. Most reports of US deployment in the setting of paediatric trauma concern abdominal imaging. It is not in common usage for acute musculoskeletal trauma, but the literature suggests that it has enormous potential, particularly with respect to elbow injuries. The use of US in sports injuries is already well established where its ability to perform dynamic assessment complements, and sometimes replaces, imaging with MR.

References and Further Reading

1. Rogers LF (1992) Radiology of skeletal trauma. Churchill Livingstone, New York, pp 109–144
2. Luckstead EF Sr, Satran AL, Patel DR (2002) Sport injury profiles, training and rehabilitation issues in American sports. Pediatr Clin North Am 49(4):753–767
3. Oudjhane K, Newman B, Oh KS, et al (1988) Occult fractures in preschool children. J Trauma 28(6):858–860
4. Kao SC, Smith WL (1997) Skeletal injuries in the pediatric patient. Radiol Clin North Am 35(3):727–746
5. Martin TJ, Martin JS (2002) Special issues and concerns for the high school- and college-aged athletes. Pediatr Clin North Am 49(3):533–552
6. Anderson SJ (2002) Lower extremity injuries in youth sports. Pediatr Clin North Am 49(3):627–641
7. Carty H (1994) Sports injuries in children—a radiological viewpoint. Arch Dis Child 70(6):457–460
8. Long G, Cooper JR, Gibbon WW (1999) Magnetic resonance imaging of injuries in the child athlete. Clin Radiol 54(12):781–791
9. Lazovic D, Wegner U, Peters G, et al (1996) Ultrasound for diagnosis of apophyseal injuries. Knee Surg Sports Traumatol Arthrosc 3(4):234–237
10. Cain EL, Clancy WG (2001) Treatment algorithm for osteochondral injuries of the knee. Clin Sports Med 20(2):321–342
11. Durston W, Swartzentruber R (2000) Ultrasound guided reduction of pediatric forearm fractures in the ED. Am J Emerg Med 18(1):72–77
12. Williamson D, Watura R, Cobby M (2000) Ultrasound imaging of forearm fractures in children: a viable alternative? J Accid Emerg Med 17(1):22–24
13. Lazar RD, Waters PM, Jaramillo D (1998) The use of ultrasonography in the diagnosis of occult fracture of the radial neck. A case report. J Bone Joint Surg Am 80(9):1361–1364
14. Hubner U, Schlicht W, Outzen S, et al (2000) Ultrasound in the diagnosis of fractures in children. J Bone Joint Surg Br 82(8):1170–1173
15. Brown J, Eustace S (1997) Neonatal transphyseal supracondylar fracture detected by ultrasound. Pediatr Emerg Care 13(6):410–412
16. Vocke-Hell AK, Schmid A (2001) Sonographic differentiation of stable and unstable lateral condyle fractures of the humerus in children. J Pediatr Orthop B 10(2):138–141
17. Markowitz RI, Davidson RS, Harty MP, et al (1992) Sonography of the elbow in infants and children. AJR Am J Roentgenol 159(4):829–833
18. May DA, Disler DG, Jones EA, et al (2000) Using sonography to diagnose an unossified medial epicondyle avulsion in a child. AJR Am J Roentgenol 174(4):1115–1117
19. Bellah R (2001) Ultrasound in pediatric musculoskeletal disease: techniques and applications. Radiol Clin North Am 39(4):597–618, ix
20. Graham DD Jr (2002) Ultrasound in the emergency department: detection of wooden foreign bodies in the soft tissues. J Emerg Med 22(1):75–79
21. Leung A, Patton A, Navoy J, et al (1998) Intraoperative sonography-guided removal of radiolucent foreign bodies. J Pediatr Orthop 18(2):259–261
22. Boutin RD, Fritz RC, Steinbach LS (2002) Imaging of sports-related muscle injuries. Radiol Clin North Am 40(2):333–362, vii
23. Gomez JE (2002) Upper extremity injuries in youth sports. Pediatr Clin North Am 49(3):593–626, vi–vii
24. Reeder MT, Dick BH, Atkins JK, et al (1996) Stress fractures. Current concepts of diagnosis and treatment. Sports Med 22(3):198–212
25. Orava S, Jormakka E, Hulkko A (1981) Stress fractures in young athletes. Arch Orthop Trauma Surg 98(4):271–274
26. Spitz DJ, Newberg AH (2002) Imaging of stress fractures in the athlete. Radiol Clin North Am 40(2):313–331
27. Anderson MW, Greenspan A (1996) Stress fractures. Radiology 199(1):1–12
28. Harty MP (2001) Imaging of pediatric foot disorders. Radiol Clin North Am 39(4):733–748
29. Sofka CM, Potter HG (2002) Imaging of elbow injuries in the child and adult athlete. Radiol Clin North Am 40(2):251–265
30. Takahara M, Ogino T, Takagi M, et al (2000) Natural pro-

gression of osteochondritis dissecans of the humeral capitellum: initial observations. Radiology 216(1):207–212
31. Lanning P, Heikkinen E (1991) Ultrasonic features of the Osgood-Schlatter lesion. J Pediatr Orthop 11(4):538–540
32. Fritz RC (1999) MR imaging of sports injuries of the elbow. Magn Reson Imaging Clin N Am 7(1):51–72, viii
33. Connolly SA, Connolly LP, Jaramillo D (2001) Imaging of sports injuries in children and adolescents. Radiol Clin North Am 39(4):773–790
34. Karasick D, Schweitzer ME (1996) The os trigonum syndrome: imaging features. Am J Roentgenol 166:125–129

3 Ultrasonography of Tendons and Ligaments

Maura Valle, Stefano Bianchi, Paolo Tomà and Carlo Martinoli

CONTENTS

3.1 Introduction 39
3.2 Normal Anatomy 39
3.2.1 Tendons 39
3.2.2 Ligaments 40
3.3 Examination Techniques and Normal Imaging Findings 40
3.3.1 US 40
3.3.2 MR imaging 42
3.4 Tendon Abnormalities 43
3.4.1 Overuse Injuries 43
3.4.2 Avulsion Injuries 43
3.4.3 Snapping Hip 46
3.4.4 Degenerative and Inflammatory Conditions 47
3.5 Ligament Abnormalities 50
3.6 Conclusion 51
References and Further Reading 51

3.1 Introduction

Magnetic resonance (MR) imaging has become established as an essential cross-sectional imaging technique for the examination of children with disorders of the musculoskeletal system. However, recent advances in ultrasound (US) technology have substantially enhanced the role of this technique to detect, localize and characterize a variety of disorders affecting tendons and ligaments in children. Although only early work is currently available in the literature on this subject, the applications of this method are maturing, and sonography is becoming the primary imaging technique for the detection, localization and characterization of a variety of tendon and ligament disorders in infants, children and adolescents.

The aim of this chapter is to describe the value of US and MR imaging in children and adolescents with a variety of diseases affecting tendons and ligaments.

M. Valle, MD
P. Tomà, MD
Reparto di Radiologia, Istituto "Giannina Gaslini", Largo Gaslini 5, 16148 Genoa, Italy
S. Bianchi, MD
Institut de Radiologie, Clinique et Fondation des Grangettes, 7, ch. des Grangettes, 1224 Chene-Bougeries, Switzerland
C. Martinoli, MD
Cattedra di Radiologia "R", Università di Genova, Largo Rosanna Benzi 8, 16132 Genoa, Italy

3.2 Normal Anatomy

3.2.1 Tendons

Tendons are structures joining the muscles to bones that allow joint movement or the maintenance of a fixed position against a loading force. There are two types of tendon, type 1 and type 2.

Type 1 tendons are long and cross one or more joints before reaching their insertions. They can reflect over bony surfaces (bony grooves or protuberances), fibrous bands or osteofibrous tunnels, and at these locations they are always surrounded by a synovial sheath made of a combination of visceral and parietal layers. The visceral layer is tightly attached to the outer tendon surface and moves with the tendon during isotonic contraction of the muscle. The parietal layer is a lax structure that surrounds the visceral synovium and blends with it at the periphery of the sheath to form the mesotendon. The main functions of the synovial sheath are to diminish friction between the tendons and the surrounding structures, thus allowing easy and smooth gliding in all positions of the adjacent joint. The sheath also forms the mesotendon that houses tendon vessels. A thin film of synovial fluid is normally found inside the tendon sheath and this may be seen in certain locations using US. For example, synovial fluid can rarely be demonstrated around the flexor digitorum tendons of the fingers, while a small amount of fluid can usually be detected in the tendon sheath of the tibialis posterior tendon and should be regarded as a normal finding.

Tendon sheaths may sometimes communicate with the adjacent articular synovial cavities. Under normal conditions a communication is present between the ankle joint and the medial tendons (tibialis posterior, flexor digitorum communis and flexor hallucis longus tendons). Therefore, excess fluid within these tendon sheaths associated with an ankle joint effusion is not necessarily the result of disease of the tendon; it may be due to leakage of fluid from an abnormal joint to a normal tendon sheath. On the other hand, some synovial sheaths do not communicate normally with the adjacent joints and even a small effusion inside them must be regarded as abnormal.

Typical examples of type 1 tendons are the flexor and extensor digitorum tendons of the hand and the ankle tendons. Due to their anatomy and the tearing forces that can develop during loading, these tendons are prone to develop friction changes and eventually partial or complete tears. Because the sheath of type 1 tendons is covered by synovium, they are commonly involved by systemic disorders that produce synovitis such as juvenile rheumatoid and seronegative arthritides.

Type 2 tendons are thicker, have a straight course and lack a synovial sheath. The paratendon is an outer envelope comprising two connective layers separated by a small amount of loose connective tissue which surrounds these tendons allowing a gliding plane with the surrounding tissue. Examples of type 2 tendons are the Achilles tendon and the quadriceps tendon.

Both types of tendon are formed by densely packed bundles of collagen fibres (type I collagen). These bundles are invested by the endotendineum and peritendineum, a network of loose connective tissue septa containing elastic fibres and vessels, which give some flexibility to the tendons. Endotendineum septa are in continuity with the epitendineum, a dense connective tissue layer tightly bound to the tendon surface.

3.2.2 Ligaments

Ligaments are flattened or cord-like periarticular structures which join two or more articular bone ends. Some ligaments, such as the anterior shoulder ligaments, are embedded in the joint capsule and cannot be differentiated from it. Other ligaments, such as the lateral collateral ligament of the knee, lie in a more peripheral location and have no relationship with the capsule. Some ligaments are formed by a single bundle of fibres, for example the anterior talofibular ligament. Others, such as the anterior cruciate ligament of the knee, are composed of multiple bundles which are subjected to different degrees of tension depending on joint position. The primary function of ligaments is to counteract excessive articular excursion, thus preventing joint subluxation and dislocation. They also maintain the position of the articular ends in the optimum alignment during movement, limiting wear and preventing early osteoarthritis. As ligaments contain a few elastic fibres scattered amongst the more resilient collagen fibres, they are slightly elastic and allow minor stretching.

3.3 Examination Techniques and Normal Imaging Findings

3.3.1 US

US examination of tendons and ligaments is best performed with high-frequency broadband (frequency range 5–15 MHz) linear array transducers to obtain a very high spatial resolution in the near field. These structures are mostly located near to the skin surface and in children they are inevitably smaller than in adults. When available small-footprint transducers are preferred as a large field-of-view is rarely required. In addition, small-sized transducers perform better around the curvature of joints and during joint or tendon motion. In infants and smaller children, large amounts of gel or a thin stand-off pad can be useful to improve the probe contact with the skin.

The sonographic appearance of tendons in children is similar to that described in adults. The main differences are due to their smaller overall size and the site of insertion into bone. When examined in the longitudinal plane, tendons appear as hyperechoic structures with well-defined echogenic (bright) margins and a fibrillar appearance due to the bundles of tendon fibres. They are anisotropic structures, which means that they may appear hypoechoic when the US beam is not precisely perpendicular to their long axis. This is because the incident US will not be reflected back to the probe unless it is exactly at 90° to the tendon fibrils. In general, anisotropy may be corrected by either examining the tendon in the position of maximal stretching or changing the orientation of the US beam by rocking the probe back and forth. Nevertheless, where tendons wind around bony surfaces and joints, for example around the ankle, anisotropy can be difficult to avoid. Its effects may be minimized only by using a careful scanning technique.

Tendons attach to a layer of hypoechoic cartilage covering the ossification centre of the bone into which they insert. The separation between tendon fibres and the ossified bone decreases with increasing patient age (Fig. 3.1). One should not misinterpret the irregular shape of the ossification centre

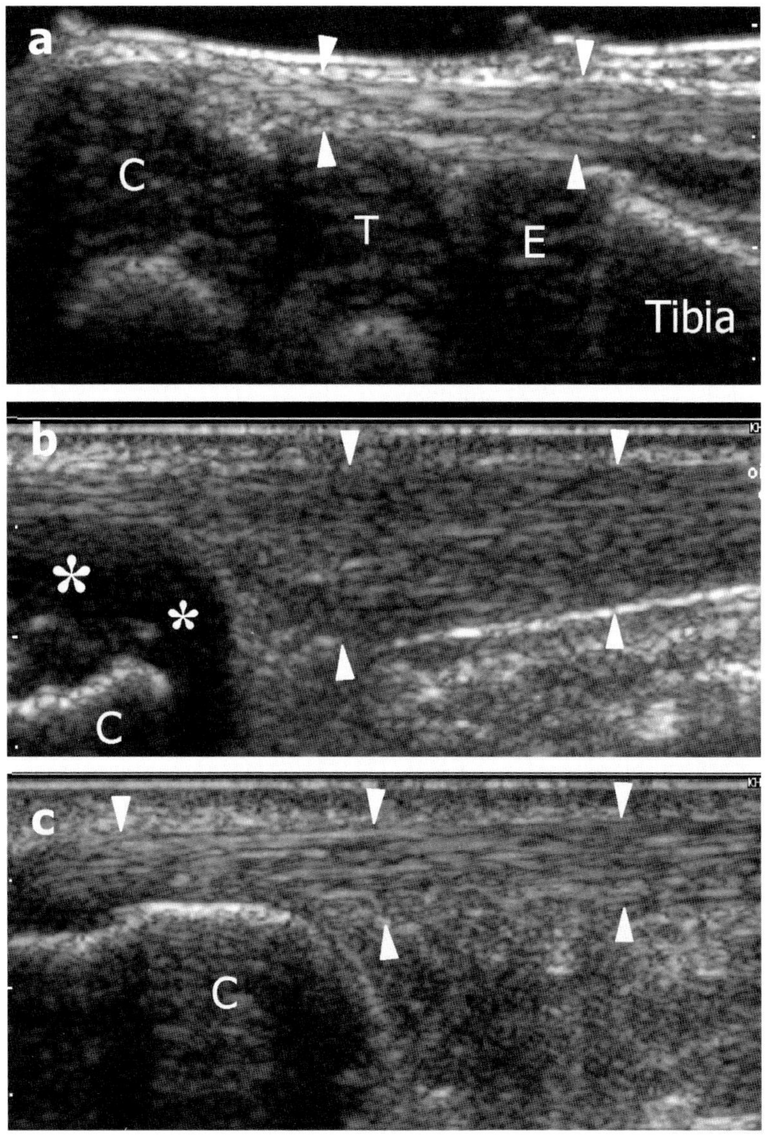

Fig. 3.1a–c. Normal US appearance of the Achilles tendon in (**a**) a 1-year-old infant, (**b**) a 5-year-old child, and (**c**) an adult. In the infant (**a**), the Achilles tendon appears as a regular hyperechoic structure (*arrowheads*) that inserts onto the posterior aspect of the calcaneus (*C*). Note that the unossified distal epiphysis of the tibia (*E*), the posterior tuberosity of the talus (*T*) and the calcaneus (*C*) are hypoechoic relative to adjacent soft-tissues, and contain fine-speckled echoes. In the child (**b**), the developing ossification centre of the calcaneus (*C*) can be appreciated as a hyperechoic structure covered by a layer of unossified cartilage (*asterisks*). The Achilles tendon is seen as it inserts onto the cartilage. In the adult (**c**), the Achilles tendon (*arrowheads*) attaches directly onto the ossified calcaneus (*C*). In all sonograms, the tendon has well-defined margins anteriorly and posteriorly and exhibits the same fibrillar echotexture made up of many parallel hyperechoic lines due to a series of specular reflections at the boundaries of collagen bundles and endotendineum septa

with the cartilaginous apophysis or epiphyses that underlie tendons as disease.

The sonographic appearance of ligaments is similar to those of tendons. Ligaments appear as hyperechoic bands with internal fibrils that join unossified hypoechoic epiphyses of adjacent bones (Fig. 3.2). Ligaments are also anisotropic structures and care should be taken not to confuse a hypoechoic area due to anisotropy with a partial tear. Bilateral examination and careful study of the ligament in different scanning planes may be helpful in avoiding misdiagnoses. Examination of ligaments should be performed at rest and during graded application of stress to the underlying joint. In selected cases, comparison images of the opposite limb may help confirm the presence of an abnormality on the symptomatic side.

3.3.2 MR Imaging

MR imaging of tendon and ligaments in children and adolescents is performed with the same protocol of pulse sequences used in adults. T1-weighted sequences (short TE/short TR) are used to obtain the best contrast resolution between tendons and ligaments and surrounding fatty tissue. Fat-suppression techniques, such as fat-saturated fast spin echo (SE) T2-weighted sequences (long TR/long TE) and fast short tau inversion recovery (fast-STIR) techniques are more effective at demonstrating structural changes, tendon sheath effusions and oedema. Compared to the fat-saturated fast SE T2-weighted sequence, fast-STIR has the advantage that it not affected by susceptibility artefacts, thus providing a more uniform fat suppression. On the other hand, the fat-suppressed fast SE T2-weighted sequence gives better anatomic definition and contrast-to-noise ratio than fast-STIR. As in adult imaging, contrast-enhanced sequences are useful in the examination of inflammatory disorders of tendons.

MR studies should be performed with the smallest coil that fits tightly around the body part being studied. In general, a flexible surface coil is better than an adult head or knee coil for examination of tendon and ligament lesions in the extremities of infants and small children. Immobilization of the limb can be achieved with a combination of tape, sponges or Velcro straps. Images are obtained in the two orthogonal planes for the structure to be

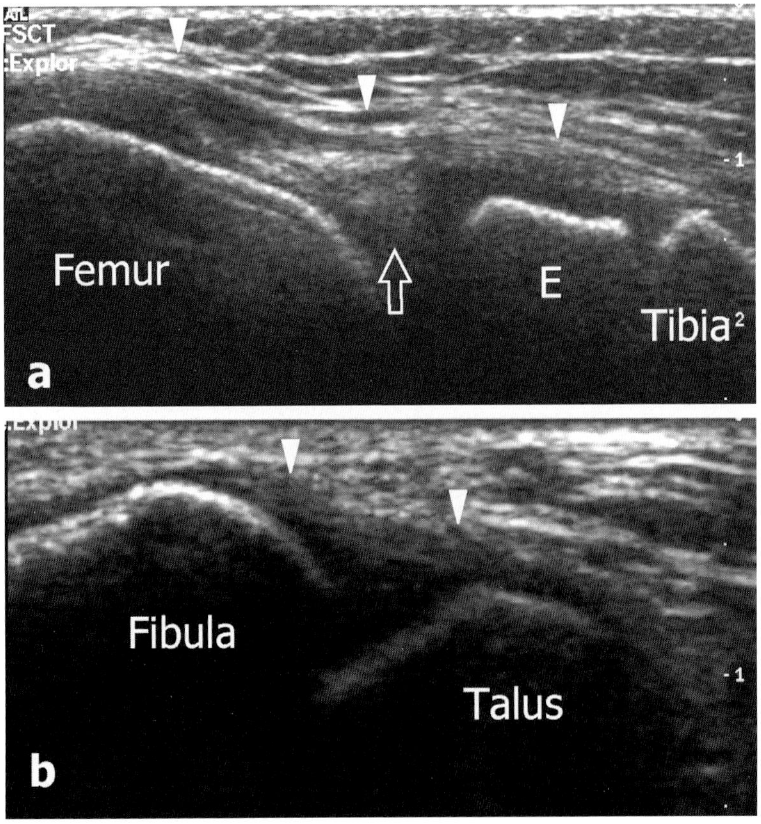

Fig. 3.2a,b. The normal US appearance of ligaments. In the knee (**a**), the medial collateral ligament (*arrowheads*) appears as a thin anisotropic band that overlies the internal aspect of the knee connecting the medial femoral condyle with the tibial epiphysis (*E*). Deep to the ligament the medial meniscus (*arrow*) appears as a hyperechoic triangular structure. In the ankle (**b**), the anterior talofibular ligament (*arrowheads*) appears as a tight hyperechoic band that joins the talus and the fibula

examined, longitudinal and axial to the tendon or ligament. High-resolution matrices (512 or 1024) and thin slices (1 to 3 mm) with minimal interslice gaps are optimum. For children of 1 year of age or younger, oral chloral hydrate (50 mg/kg) is used for sedation. When the child is older than 6 years, sedation is unnecessary in most cases. Monitoring the sedated child during the examination by staff trained in anaesthesia with equipment safe for use with MR is mandatory.

Due to the absence of internal free water, normal tendons and ligaments appear as homogeneously hypointense structures on both T1- and T2-weighted images. Some tendons with a curvilinear course may exhibit focal signal changes caused by tissue anisotropy when their fibres run at 55° with respect to the magnetic field (magic-angle effect). Examiners should be aware of this artefact to avoid confusion with disease.

3.4 Tendon Abnormalities

A variety of disorders can affect tendons in children, although they occur less commonly than in adults. The more common indications for sonography of tendons are trauma, snapping iliopsoas tendon, and degenerative, inflammatory and infectious conditions. The weakest point of the muscle–tendon–bone unit in children is not the musculotendinous junction or the tendon substance, as seen in adults, but the attachment of the tendon to the non-ossified cartilage. Therefore, most tendon disorders in children, and especially in school-aged athletes, involve the tendino-osseous junction whilst degenerative changes and ruptures in the tendon substance occur infrequently. Two main types of abnormality are observed: acute trauma that results in partial or complete detachment of the apophysis by avulsion at the site of tendon insertion, and chronic lesions when repeated microtrauma secondary to overload leads to osseous or cartilage fragmentation. The latter is often seen following repetitive activity in sports. In both circumstances, the diagnosis is based on clinical findings with a history of pain during isometric muscle contraction. Sonography is increasingly being used to confirm the clinical suspicion.

3.4.1 Overuse Injuries

Overuse injuries are the consequence of exceeding the ability of tendon insertion to recover from submaximal cyclic loading in tension, compression, shear or torsion, and depend on a variety of factors, including tissue strength, joint size, and the patient's age and skeletal maturity. Chronic traction on the apophyseal cartilage by a tendon or a ligament may result in progressive physeal microfracture, widening, avulsion of fragments of cartilage and bone and insertional tendinopathy. By far the most common site involved is the knee, with injury to the insertions of the patellar tendon, either the anterior tibial apophysis (Osgood-Schlatter disease) or the lower pole of the patella (Sinding-Larsen-Johansson disease or jumper's knee). These conditions occur in adolescents and may present with pain exacerbated by activity, local soft-tissue swelling and tenderness. Osgood-Schlatter disease usually affects boys with a history of participation in sports and a rapid growth spurt. Sinding-Larsen-Johansson disease is similar to jumper's knee. In both diseases, standard lateral radiographs can demonstrate a fragmented appearance of the apophysis. High-resolution US is an accurate means to reveal even small calcified fragments and irregularity in the bony outline resulting from the osteochondrosis. It will demonstrate focal hypoechoic swelling of the physeal cartilage, hypoechoic changes in the patellar tendon from tendinosis and fluid collection from infrapatellar bursitis (Fig. 3.3) [1–4]. In the acute phase, local hyperaemia can be demonstrated with colour and power Doppler imaging [2]. Similar to the signs observed in the knee, the posterior apophysis of the calcaneus can undergo fragmentation (Sever's disease) leading to chronic heel pain. In these patients, US is also suitable for noninvasive follow-up of the disease. MR imaging findings include increased T2-weighted signal at the insertion of the tendon, in the surrounding soft tissue and in the adjacent bone marrow [5].

3.4.2 Avulsion Injuries

Following substantial trauma, avulsion injuries may occur. The pelvic girdle is the site most frequently affected. Around the pelvis, high-resolution US is able to detect apophyseal avulsion at the ischial tuberosity (hamstrings muscles), the anterior supe-

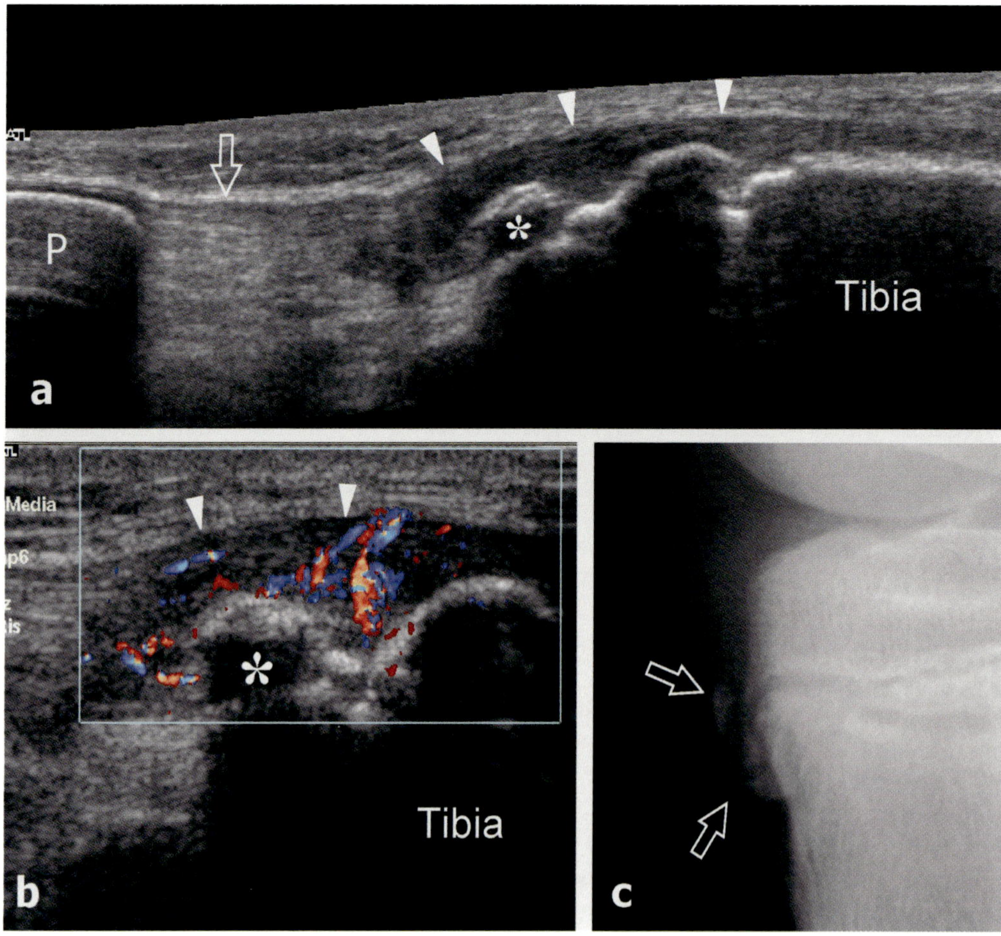

Fig. 3.3a–c Osgood-Schlatter disease. Longitudinal 12-5 MHz grey-scale (**a**) and colour Doppler (**b**) images of the patellar tendon in a 15-year-old boy with focal tenderness and chronic pain over the tibial tuberosity reveal a swollen hypoechoic distal patellar tendon (*arrowheads*) and bony irregularity and fragmentation of the anterior tibial surface (*asterisk*); *P* patella. In the colour Doppler image (**b**), local increased flow signals (*arrowheads*) reflect intratendinous hyperaemia. A lateral radiograph (**c**) demonstrates a fragmented irregular apophysis (*arrows*)

rior iliac spine (sartorius muscle and tensor fascia lata) and anterior inferior iliac spine (rectus femoris muscle), the iliac crest (abdominal and gluteus medius muscles), the lesser trochanter (iliopsoas muscle), the greater trochanter (external rotators) and the symphysis pubis (adductor muscles). At these sites, the fracture edge may extend directly through the physeal cartilage, into the ossifying apophysis or the underlying bone. US can demonstrate an irregular bony surface, a thickened physeal cartilage with fissures, small hyperechoic structures with posterior acoustic shadowing from avulsed bone fragments and local haematoma (Fig. 3.4). The size and amount of displacement of the avulsed fragment is variable. In doubtful or difficult cases, MR imaging may be a useful adjunct to US. The main advantages of this technique include better images of deep-seated tendons or difficult-to-scan regions (Fig. 3.5).

An injury that occurs at the attachment of tendons to the unossified skeleton is the sleeve fracture that commonly occurs at the poles of the patella (proximally, insertion of the quadriceps tendon; distally, insertion of the patellar tendon), the proximal olecranon and the medial epicondyle at the elbow. These injuries are due to failure of the physis following excessive traction stresses. US identifies a broad sleeve of cartilage, often associated with an osseous fragment pulled away with the tendon (Fig. 3.6) [6]. In more subtle cases of apophyseal irritation with minimal displacement, high-resolution US may demonstrate a "double cortical sign" as a result of

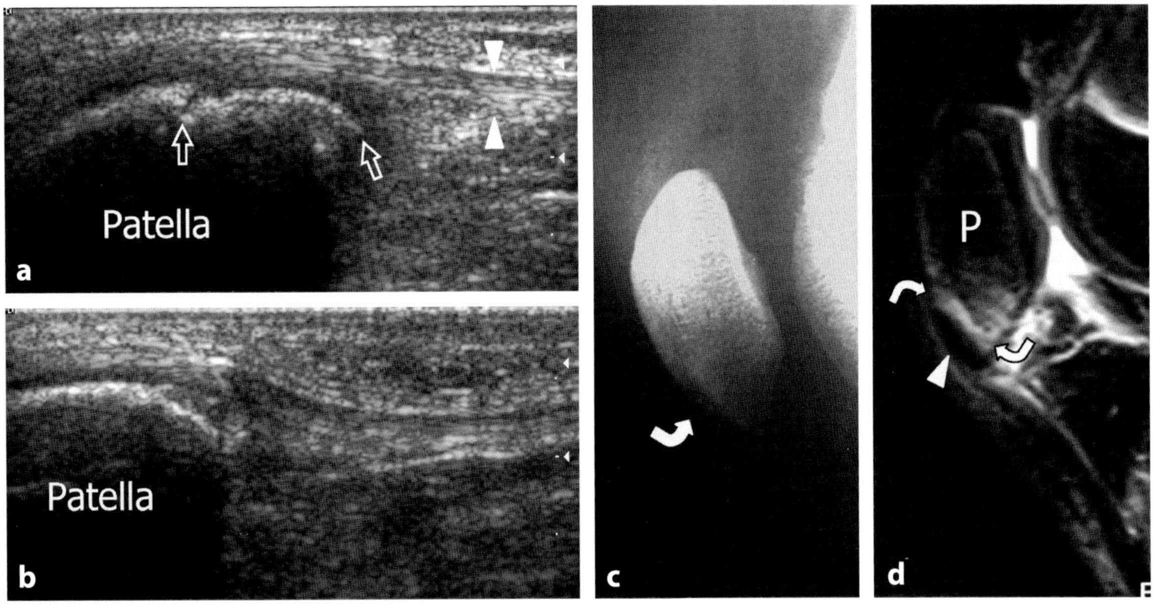

Fig. 3.4a–d. Traction injury at the lower pole of the patella of a 14-year-old boy following a kick during a soccer game. **a** Longitudinal 12-5 MHz US image over the proximal insertion of the patellar tendon (*arrowheads*) demonstrates cortical irregularity (*arrows*) of the lower pole of the patella consistent with minimal displacement of a fleck of bone. **b** Contralateral healthy side. **c** Radiograph shows a small fragment of bone (*arrows*) detached from the patella. **d** Sagittal fat-suppressed T2-weighted MR image of the patella (*P*) identifies the bony fragment (*arrowhead*) in continuity with the patellar tendon, associated with marrow oedema (*curved arrows*) at the detachment site

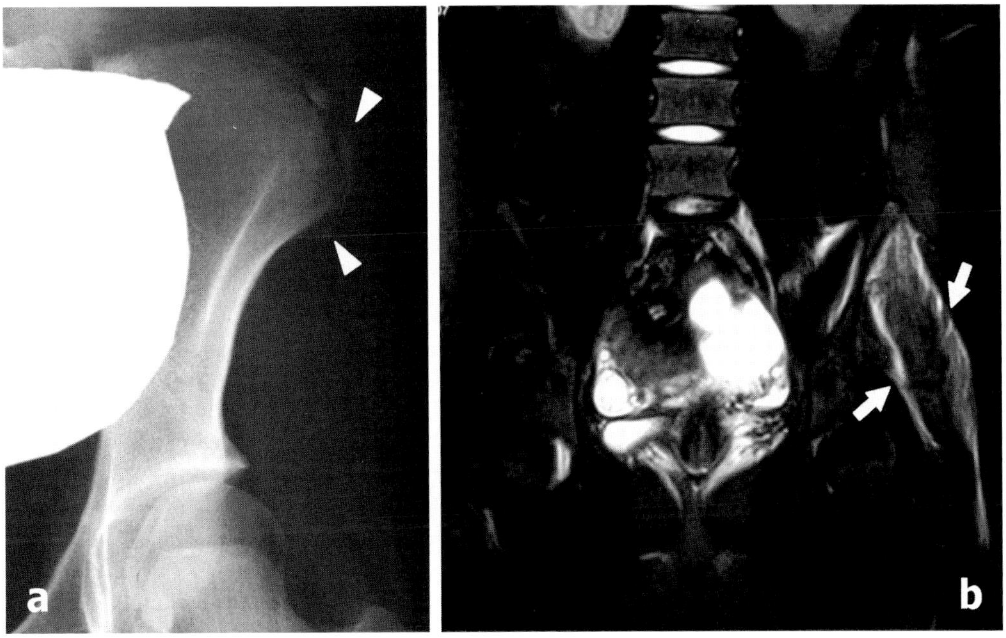

Fig. 3.5a–c. Acute avulsion of the iliac crest in a 15-year-old soccer player. **a** Radiograph shows a displaced ossification centre (*arrowheads*) of the left iliac crest. **b** Coronal fat-suppressed T2-weighted MR image of the pelvic girdle demonstrates diffuse oedematous changes in the gluteus medius muscle (*arrows*) related to the trauma

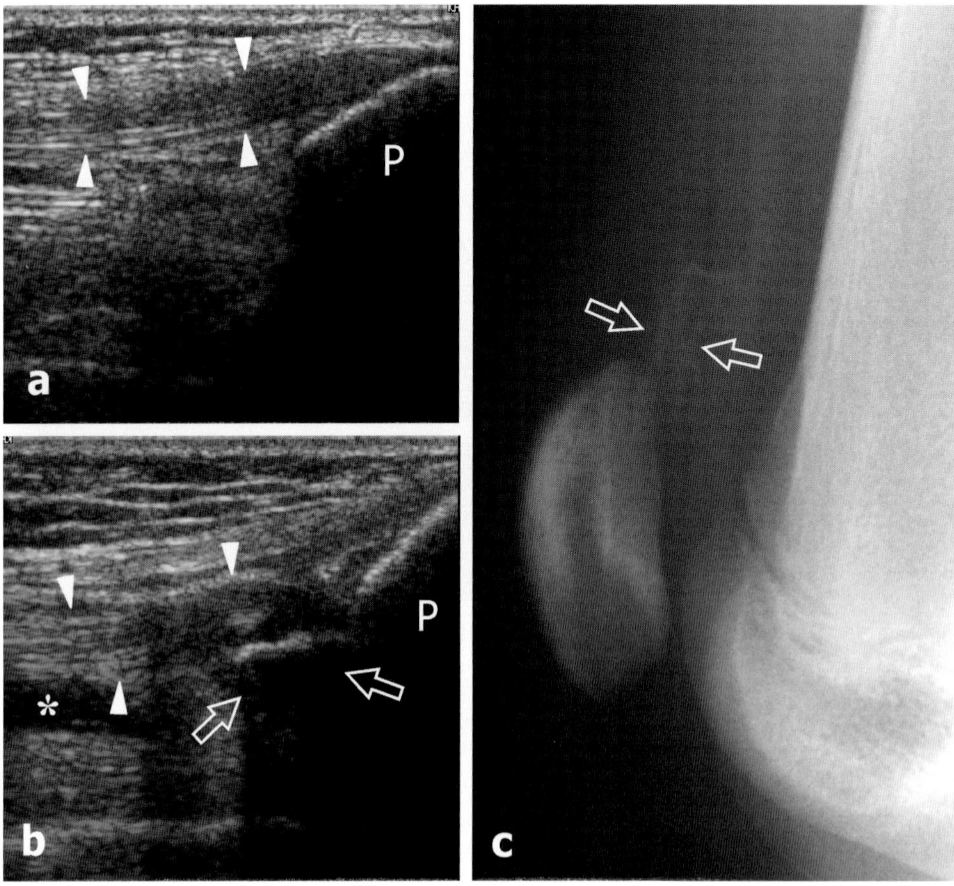

Fig. 3.6a–c. Avulsion sleeve fracture of the patella. Longitudinal 12-5 MHz US images obtained over the dorsal aspect of the distal left (**a**) and right (**b**) quadriceps tendon in a 8-year-old child with complete inability to knee extension after an acute injury. In the left quadriceps tendon (**a**), the normal contralateral tendon (*arrowheads*) shows well-defined borders and normal internal echo texture; *P* upper pole of the patella. In the right quadriceps tendon (**b**), the affected quadriceps tendon (*arrowheads*) appears swollen and hypoechoic. The tendon attaches to a hyperechoic bony structure (*arrows*) that lies deep and cranial to the upper pole of the patella (*P*). This finding indicates a posttraumatic avulsion injury at the upper pole of the patella. Note the intra-articular effusion located inside the suprapatellar synovial pouch (*asterisk*). **c** A standard radiograph confirms the presence of an avulsed osseous fragment (*arrows*) from the patella

elevation of the most superficial layer of the bony cortex (Fig. 3.7) or a wavy and thickened periosteal line, separated from the bone by an effusion. When a traction injury is strongly suspected on clinical grounds and US is negative, MR imaging is the study of choice to identify the lesion by observing marrow oedema with widening and irregularity of the physis. The degree of fragment displacement is critical in therapeutic planning. Most cases will require surgery with the possible exception of those with minimal displacement.

3.4.3 Snapping Hip

A tendon abnormality that may be encountered in the adolescent is the so-called "snapping hip". This disease is often bilateral and presents with an audible snap produced during walking or hip movement. It is due to snapping of either the iliopsoas tendon over the iliopectineal eminence or the iliotibial band over the greater trochanter (Fig. 3.8). The painful snapping hip has been described in adolescents (average age, 15 years) involved in competitive athletic activities, and rarely in association with habitual hip dislocation [7, 8]. Dynamic US is an ideal means to identify this condition by showing the iliopsoas

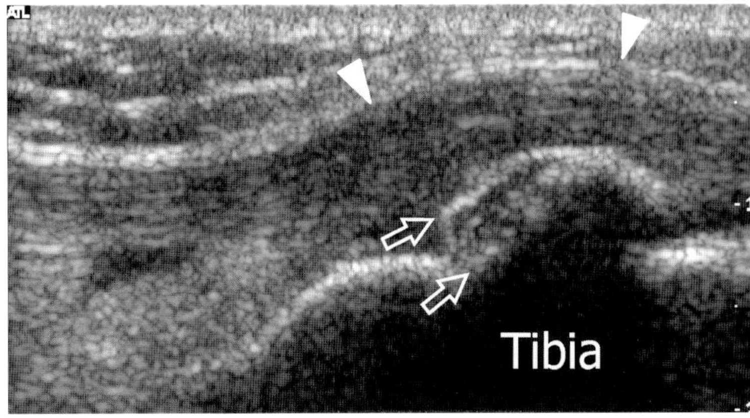

Fig. 3.7. Double cortical sign in a 14-year-old sprinter with a recent acute traction trauma and pain over the tibial tuberosity. Longitudinal 12-5 MHz US image shows a thickened patellar tendon (*arrowheads*) and the elevation and fragmentation of the cortical bone of the tibial tuberosity forming two hyperechoic layers (*arrows*)

tendon or the iliotibial band which suddenly display an abrupt abnormal lateral displacement during hip movement [9, 10]. Conservative treatment with rest and antiinflammatory drugs is sufficient in most patients. In cases of instability of the iliopsoas tendon, surgical lengthening of the tendon may be needed in cases with persistent pain.

3.4.4
Degenerative and Inflammatory Conditions

Degenerative disorders of tendons are rare in children and usually follow mechanical stress related to foot disorders, including clubfoot and flat foot (Fig. 3.9). In chronic renal failure treated by haemodialysis, amyloid deposits can be seen both in the paraarticular tissues and within the tendon substance. The amyloid deposits cause swelling of the involved tendon and a more heterogeneous appearance of the fibrillar echo texture. Occasionally, a hyperaemic pattern can be found at colour and power Doppler examination.

Differing from traumatic and degenerative lesions, the inflammatory involvement of tendons invested by synovial sheath is commonly encountered in patients with juvenile idiopathic arthritis. The US appearance of the affected tendons varies depending on the stage of synovial involvement (acute vs chronic). In the early stages, the tendon has a normal size and echotexture and is surrounded by an anechoic area produced by the synovial effusion. In more advanced disease, synovial hypertrophy

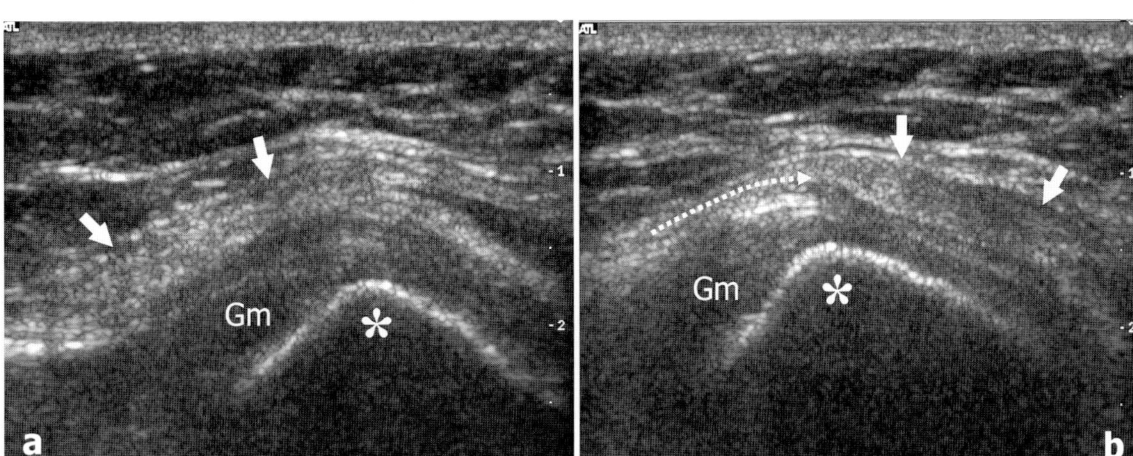

Fig. 3.8a,b. Snapping iliopsoas band. Transverse 12-5 MHz US images of the lateral aspect of the right hip. When the hip is flexed (**a**), the iliotibial band is present as a hyperechoic stripe (*arrows*) posterior to the trochanter (*asterisk*) and superficial to the gluteus medius tendon (*Gm*). During extension of the hip (**b**), an abrupt displacement (*dotted arrow*) of the iliotibial band occurs as it gets closer to the trochanter, coinciding with the snapping sensation

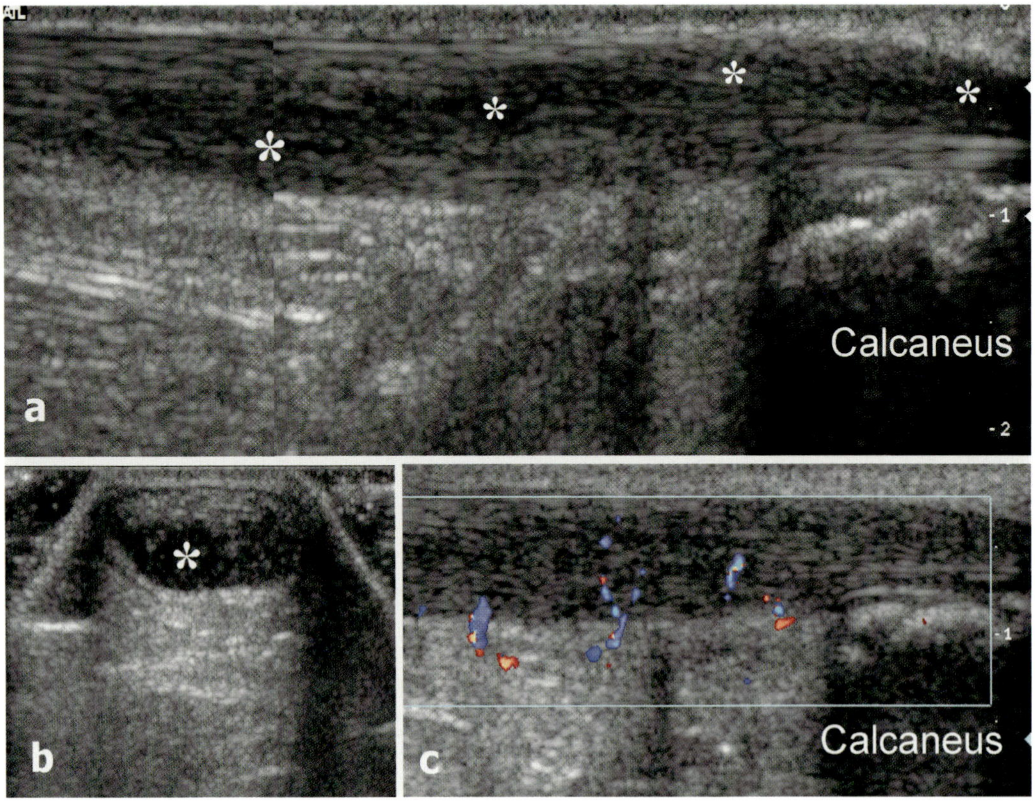

Fig. 3.9a–c. Degenerative changes in the Achilles tendon of a 10-year-old boy who was previously operated upon for flat foot. The patient presented with chronic heel pain. Longitudinal (**a**) and transverse (**b**) grey-scale 12-5 MHz US images obtained over the Achilles tendon demonstrate diffuse fusiform hypoechoic swelling (*asterisks*) of the tendon extending from its insertion to approximately 3 cm above the calcaneus due to microtears and mucoid degeneration. The colour Doppler image (**c**) shows an increased depiction of intratendinous flow signals. The pattern distribution of flow is characterized by vessel pedicles that enter the tendon from its anterior surface

can be seen as hypoechoic folds projecting inside filling the synovial sheath (Fig. 3.10). In active tenosynovitis, colour and power Doppler imaging and gadolinium-enhanced MR sequences can aid differentiation between pannus and effusion by showing flow signals inside the synovium (Fig. 3.11) or contrast enhancement. However, one should always keep in mind that the fibrous pannus does not show hypervascular changes. A definite advantage of US is the ability to differentiate tendon involvement from arthritis of the underlying joints (Fig. 3.12).

These conditions require different treatments and may be difficult to discriminate on physical examination. In longstanding disease, the involved tendons may become swollen and hypoechoic. In treated juvenile idiopathic arthritis, the occurrence of partial and complete tendon tears is rare, mostly confined to hand and ankle tendons. In partial tendon tears, US demonstrates focal swelling or thinning of the involved tendon, whereas in complete tears a gap is observed between hypoechoic and oedematous tendon ends. Although the diagnosis of complete tears is usually straightforward on clinical examination, high-resolution US can help the surgeon assess the amount of retraction of the proximal tendon end as well as plan an adequate skin incision. In specific clinical settings, US can provide an accurate and confident guidance to direct the needle for synovial biopsy procedures and for the injection of corticosteroids inside the tendon sheath.

The differential diagnosis of inflammatory tenosynovitis includes infection. If a synovial sheath tendon is involved in isolation, the possibility of an infectious tenosynovitis should be considered, especially in cases with a history of a penetrating injury. In patients with juvenile idiopathic arthritis, distinguishing infectious from arthritic involvement may be challenging. Most of these patients are being treated with corticosteroids which may mask the signs of infection (fever, pain, limitation of move-

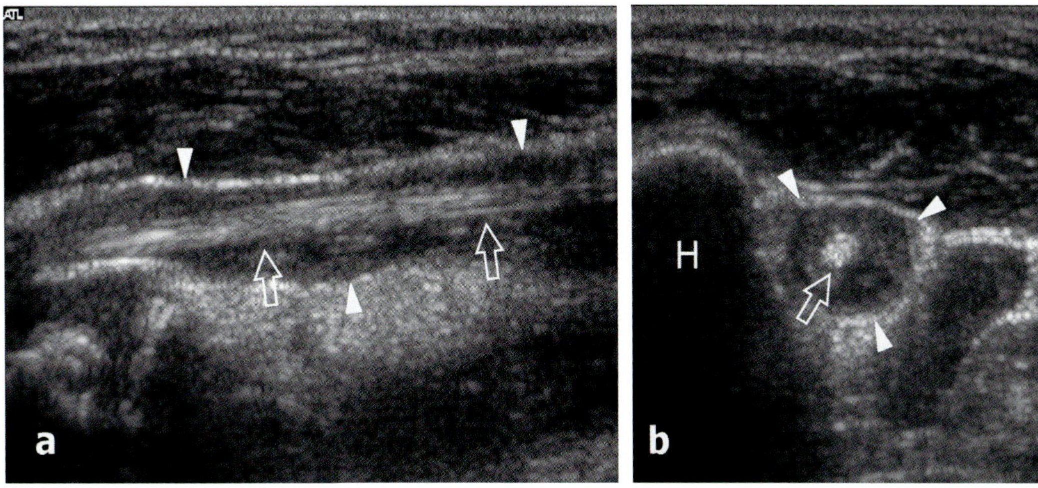

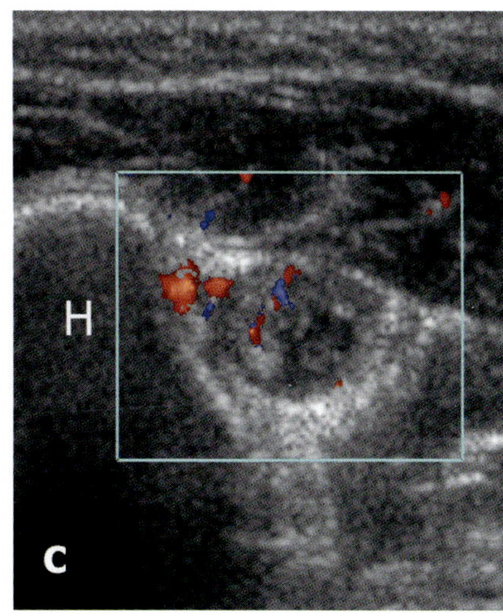

Fig. 3.10a–c. Tenosynovitis of the long head of the biceps tendon in a 5-year-old child with juvenile idiopathic arthritis. Longitudinal (**a**) and transverse (**b**) 12-5 MHz US images over the anterior aspect of the shoulder demonstrate thickening of synovial tissue (*arrowheads*) around the long head of the biceps tendon (*arrows*) related to synovial hypertrophy. The patient had also distension of the anterior recess of the glenohumeral joint. Colour Doppler imaging (**c**) shows hypervascularity of the synovial sheath as a sign of active synovial pannus; *H* humerus

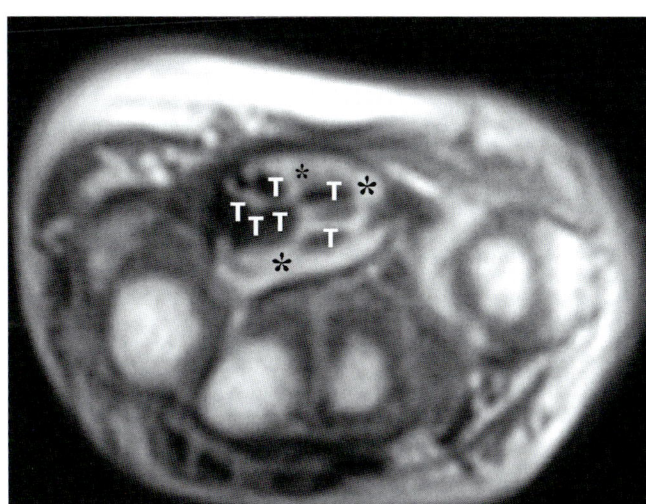

Fig. 3.11. Flexor tenosynovitis of the wrist in an 8-year-old female with juvenile rheumatoid arthritis. Axial gadolinium DTPA-enhanced T1-weighted MR image of the carpal tunnel demonstrates contrast uptake of the synovial pannus (*asterisks*) surrounding the flexor tendons (*T*)

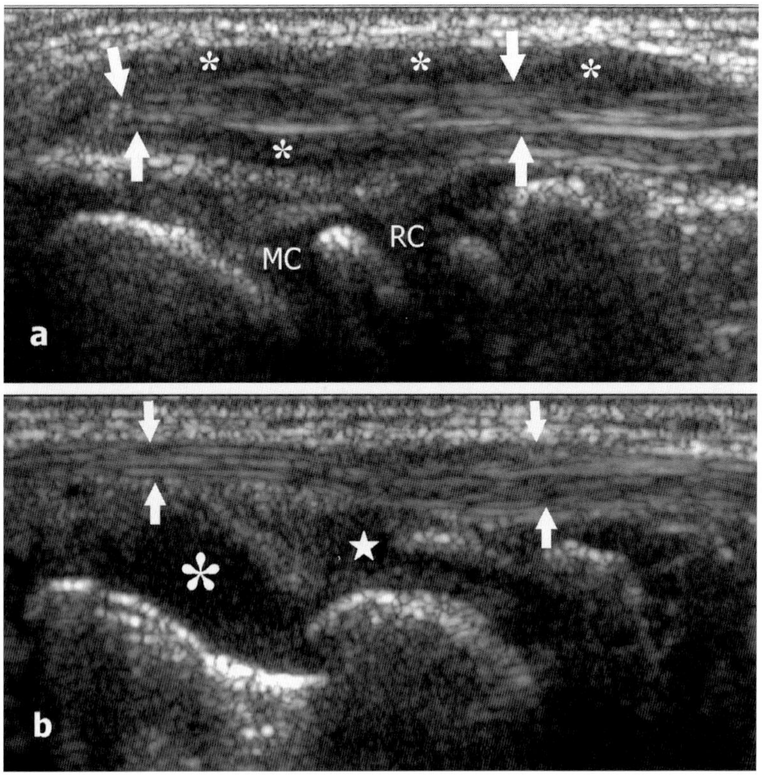

Fig. 3.12a,b. Differentiation between tenosynovitis and joint synovitis. Longitudinal 12-5 MHz US images obtained over the dorsal aspect of the wrist in two different patients affected by juvenile idiopathic arthritis. **a** The extensor tendons (*arrows*) are thickened and surrounded by hypoechoic synovial pannus (*asterisks*). Deep to these tendons, the dorsal recesses of the radiocarpal (*RC*) and mediocarpal (*MC*) joints appear normal. **b** The extensor tendons (*arrows*) are normal, without any process involving their sheaths. Instead, a definite effusion is detected within radiocarpal (*small asterisk*) and mediocarpal (*large asterisk*) joints indicating joint synovitis

ment). Sonography does not allow a reliable differentiation between these conditions, but it can guide the needle aspiration of the sheath fluid for Gram staining and culture.

3.5 Ligament Abnormalities

Ligament injuries are rare in children. They are almost exclusively observed in adolescents involved in competitive sports, but they may be the result of trauma. There is a paucity of literature on the application of sonography to ligament imaging in children. High-resolution US has proved to be able to reveal injuries of the anterior talofibular and anterior tibiofibular ligaments in children with ankle injuries, although unexpectedly, sonography identifies more ligament injuries than physeal injuries in these patients [11]. In partial-thickness tears, US demonstrates an irregular swelling of the severed ligament which retains its straight appearance. Focal hypoechoic areas are more often seen at its proximal and distal attachments. Chronic partial tears may exhibit calcification and scar tissue (Fig. 3.13). In full-thickness tears, the ligament is interrupted and its ends are separated by either hypoechoic blood collection (acute) or hyperechoic fibrosis (chronic). At the medial aspect of the elbow, the anterior band of the medial collateral ligament complex may be injured in adolescents who practice baseball pitching or throwing sports. In a partial tear of this ligament, high-resolution US reveals a thickened hypoechoic and irregular ligament surrounded by effusion. In complete rupture, US may show either a gap or focal hypoechoic areas in the proximal and distal aspects of the ligament.

In ligament injuries, MR imaging is the method of choice when the involved structure is inaccessible to US examination (for example, the cruciate ligaments of the knee and other intraarticular or deeply located structures). The MR features of partial or complete ligament tears are the same as those described in adults. In particular, a partial tear produces increased signal intensity within the involved ligament with some fibres remaining intact, whereas a complete ligament tear leads to complete discontinuity of the ligament fibres and increased signal intensity that extends completely across the ligament on T1- and T2-weighted images. Alternatively there may be complete absence of the ligament.

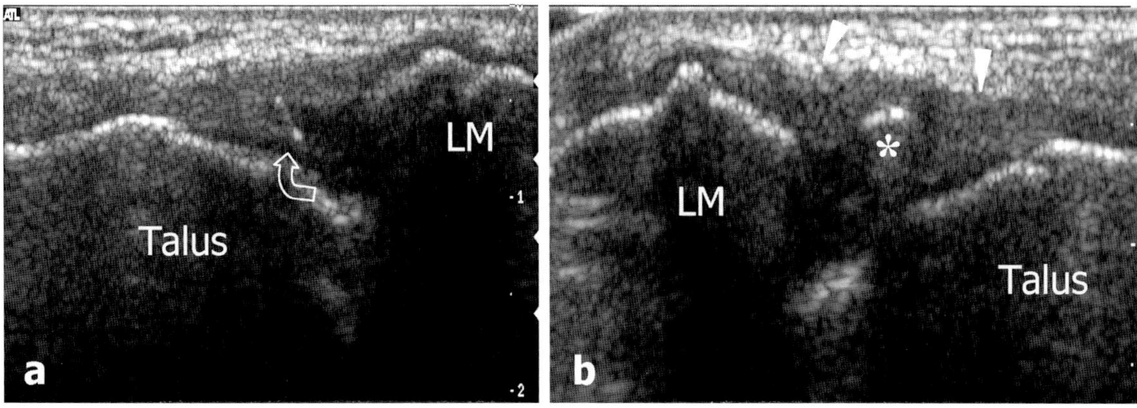

Fig. 3.13a,b Partial tear of the right anterior talofibular ligament of the ankle in a 15-year-old girl involved in agonistic athletic activity. **a** Longitudinal 12-5 MHz US image over the right anterior talofibular ligament demonstrates discontinuity (*curved arrow*) of its deep fibres. **b** The normal left ligament exhibits a hyperechoic focus (*asterisk*) reflecting posttraumatic calcification; *LM* lateral malleolus

3.6 Conclusion

Recently improved performance of "small parts" transducers has enabled US to demonstrate a greater range of abnormalities that affect tendons and ligaments of children and adolescents. Although MR imaging offers superior soft-tissue contrast resolution, US is low-cost, noninvasive, of higher spatial resolution and provides real-time capability for the assessment of these structures during joint movements and stress manoeuvres. We believe that US should be regarded as the first-line imaging technique in the care of patients thought to be suffering from tendon and ligament disorders. It is our view that this approach will avoid the need for MR in many cases. MR examination should be reserved for patients with complex injuries, such as osteochondral and avulsion injuries where the cartilaginous or bony component is an important part of the lesion. It is useful to define the extent of the injury, determine joint involvement and assess fragment displacement. Radiologists must be able to apply both imaging methods when investigating tendon and ligament disorders in children.

References and Further Reading

1. Blankstein A, Cohen I, Heim M, et al (2001) Ultrasonography as a diagnostic modality in Osgood-Schlatter disease. A clinical study and review of the literature. Arch Orthop Trauma Surg 121:536–539
2. Carr JC, Hanly S, Griffin J, et al (2001) Sonography of the patellar tendon and adjacent structures in pediatric and adult patients. AJR Am J Roentgenol 176:1535–1539
3. De Flaviis L, Nessi R, Scaglione P, et al (1989) Ultrasonic diagnosis of Osgood-Schlatter and Sinding-Larsen-Johansson diseases of the knee. Skeletal Radiol 18:193–197
4. Lanning P, Heikkinen E (1991) Ultrasonic features of the Osgood-Schlatter lesion. J Pediatr Orthop 11:538–540
5. Rosenberg ZS, Kawelblum M, Cheung YY, et al (1992) Osgood-Schlatter lesion: scintigraphic, CT and MR imaging features. Radiology 185:853–858
6. Bates DG, Hresko MT, Jaramillo D (1994) Patellar sleeve fracture: demonstration with MR imaging. Radiology 193:825–827
7. Dobbs MB, Gordon JE, Luhmann SJ, et al (2002) Surgical correction of the snapping iliopsoas tendon in adolescents. J Bone Joint Surg Am 84:420–424
8. Walker J, Rang M (1992) Habitual hip dislocation in a child: another cause of snapping hip. Clin Pediatr 31:562–563
9. Choi YS, Lee SM, Song BY, et al (2002) Dynamic sonography of external snapping hip syndrome. J Ultrasound Med 21:753–758
10. Pelsser V, Cardinal E, Hobden R, et al (2001) Extraarticular snapping hip: sonographic findings. AJR Am J Roentgenol 176:67–73
11. Farley FA, Kuhns L, Jacobson JA, et al (2001) Ultrasound examination of ankle injuries in children. J Pediatr Orthop 21:604–607

4 Inflammatory Disorders

David Wilson

CONTENTS

4.1 Introduction 53
4.2 Irritable Hip 53
4.2.1 Clinical Background 53
4.2.2 Role of Imaging in Detection 54
4.2.2.1 Plain Films 54
4.2.2.2 Nuclear Medicine 54
4.2.2.3 US 54
4.2.2.4 CT 56
4.2.2.5 MRI 56
4.2.3 Role of Imaging in Treatment 57
4.2.4 Management Protocol 57
4.2.5 Potential Developments 58
4.3 Synovitis 58
4.3.1 Clinical Background 58
4.3.2 Role of Imaging 58
4.3.3 Potential Developments 59
4.4 Osteomyelitis and Soft-tissue Infection 59
4.4.1 Clinical Background 59
4.4.2 Role of Imaging in Detection 60
4.4.3 Role of Imaging in Treatment 64
4.4.4 Potential Developments 64
References and Further Reading 64

4.1 Introduction

Inflammatory conditions of joints normally present with pain, swelling and dysfunction. Children over the age of three to four years of age with joint pain will usually tell their parents that this is the problem. However, the diagnosis is not so easy in infants for whom the presentation may be that the parents notice that they are not using the limb, have stopped walking or even, less specifically, have gone off their food or have altered in mood inexplicably. Careful physical examination may be required to locate the problem. The fractious and unwell child may resist this examination and it can prove very difficult to make or exclude acute arthralgia. Other diseases may mimic joint pain. Muscle injury is a common mimic and the history is rarely helpful. All young children are involved in incidents that might pull muscles regularly and few can relate an injury to a specific event.

Imaging has a number of potential roles. Principally these are:

- To confirm that a joint is inflamed by detecting an effusion
- To demonstrate that a joint is normal, directing attention to other possibilities
- To define or exclude adjacent bone involvement
- To guide aspiration of a joint for diagnostic and therapeutic purposes
- To plot the joints affected by systemic arthropathies

Any joint may be affected by infection, transient synovitis and systemic arthropathy. However, the hip is a common site and serves to illustrate the range of possible diagnoses and the potential assistance afforded by imaging. The same principles apply to all the other joints of the body.

4.2 Irritable Hip

4.2.1 Clinical Background

Irritable hip is the clinical syndrome that most commonly affects children between the age of four and ten years. It is most often due to transient synovitis which is a self-limiting condition for which no cause has been found. However, there are many other potential causes some of which require urgent medical attention if serious consequences are to be

D. Wilson, FRCP, FRCR
Department of Radiology, Nuffield Orthopaedic Centre, NHS Trust, Windmill Road, Headington, Oxford, OX3 7LD, UK

avoided. The list of possible causes of an irritable hip includes:

- Transient synovitis
- Muscle or tendon injury
- Septic arthritis
- Perthes' disease
- SUFE
- Systemic arthropathy
- Crystal synovitis
- Haemorrhagic diatheses

There may be clues in the history that assist—for example, previous episodes, family predisposition or a tendency to bleed easily. However, there is so much overlap it is unwise to rely on the clinical presentation to exclude the more serious causes of joint pain. Studies have shown that some children with aggressive septic arthritis do not exhibit clinical features that would discriminate them reliably from transient synovitis. Fever, intensity of pain, duration of pain, ESR (erythrocyte sedimentation rate), CRP (C reactive protein) and general malaise have all been studied and many, if not all, may be normal or absent despite the presence of a pyogenic organism in the joint. It is true that these features are more common in septic arthritis [1] but as we only see one patient at a time it is potentially dangerous to stop or delay investigation because the clinician is "sure that it is not infected".

4.2.2 Role of Imaging in Detection

An effusion in relatively superficial joints such as the knee is often straightforward to detect clinically. However, in the shoulder or hip, large effusions may be occult clinically and in complex joints such as the wrist it may be difficult to distinguish ganglia from effusion and synovitis. Indeed, even in the knee a small collection may be overlooked and it is recognized that the amount of fluid does not predict the severity. Pyogenic septic arthritis may present with a small amount of fluid in the joint.

4.2.2.1 Plain Films

Some authors and textbooks suggest that displacement of fat planes may indicate an effusion. This is true, but potentially misleading. In the knee displacement of the fat and muscle planes around the suprapatellar pouch is sensitive to small effusions, although it is likely that clinical examination is just as sensitive. In the elbow, provided that the image is a true lateral, displacement of fat pads is an excellent and reliable means of detecting or excluding an effusion. However, in the shoulder there a few planes that help and in the hip the position of the fat/muscle interfaces is more related to external rotation than being a true sign of an effusion. In 1966 Kemp and Boldero described lateral displacement of the femoral head associated with effusion in the hip of children with early Perthes' disease [2]. The authors attributed the displacement to cartilage oedema and thickening, but others who misread and misinterpreted the paper thought that the sign indicated an effusion. Recent ultrasound (US) evidence and correlation confirms the original hypothesis, but unfortunately the error has been transcribed through generations of textbooks. Suffice to say that there are no reliable signs of an effusion of the hip on plain films [3].

4.2.2.2 Nuclear Medicine

Bone scintigraphy with Tc99m MDP has been proposed as a means of detecting hip disease in children with an irritable hip. It will show increased activity on both blood pool (vascular) phase and crystal (late) phase images in almost all the causes of irritable hip listed above. There may be areas of photopenia in certain stages of early Perthes' disease but is less sensitive than MRI. Effusions are not directly detected, and a negative scintigram does not exclude hip disease. Since the advent of MRI there is now little role for scintigraphy in the management of the irritable hip [4].

4.2.2.3 US

Ultrasound is the best, fastest and most accurate means of detecting joint effusions (Figs. 4.1–4.3) [5, 6]. Most fluid will be echo-free but some effusions contain particulate matter that creates some internal echoes. When there is doubt, one should first compare the potential area of fluid with known areas of fluid at the same depth, such as a vessel or the bladder. The gain setting should then be adjusted to make the known fluid appear just echogenic then turning the gain back to the point at which it appears just echo-free, and re-examining the suspect areas with a more accurate means of typing the material in the joint. Echogenic effusions include very recent

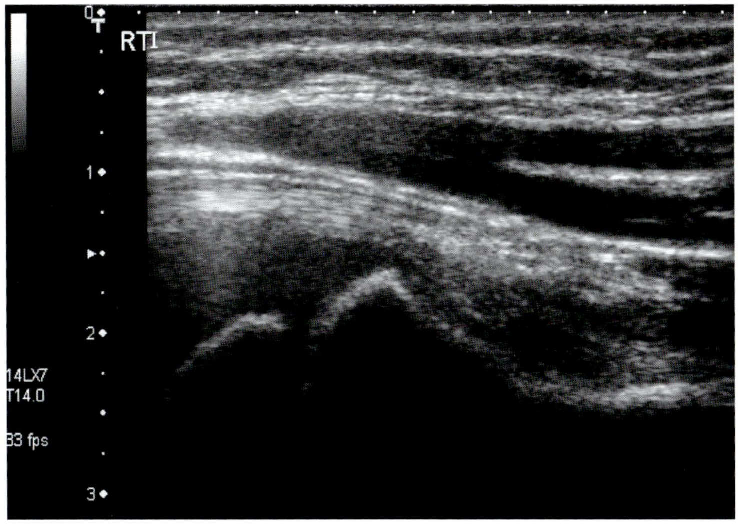

Fig. 4.1 Ultrasound of a normal hip

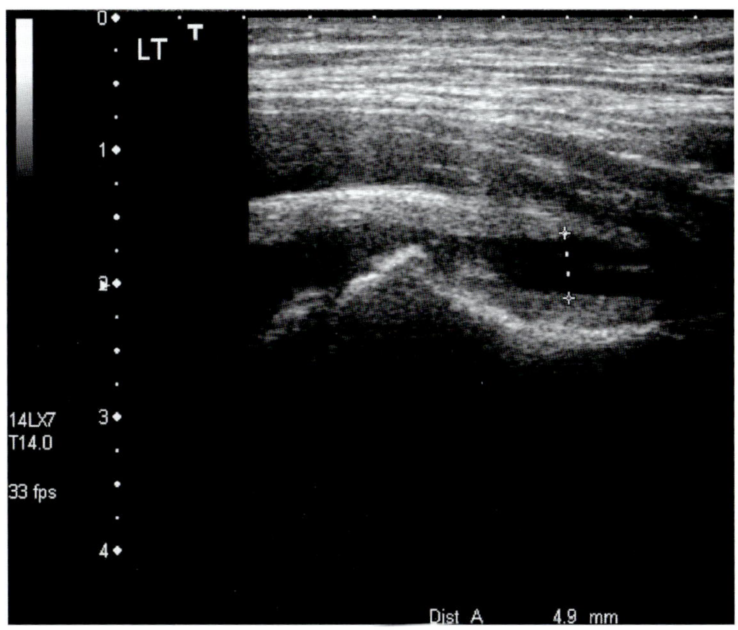

Fig. 4.2 Ultrasound of a hip with an echo-free (*black*) joint effusion. Measurements should be from reference points identifiable in both hips in the same direction with the limbs in the same position. A difference of 2 mm is significant [5]

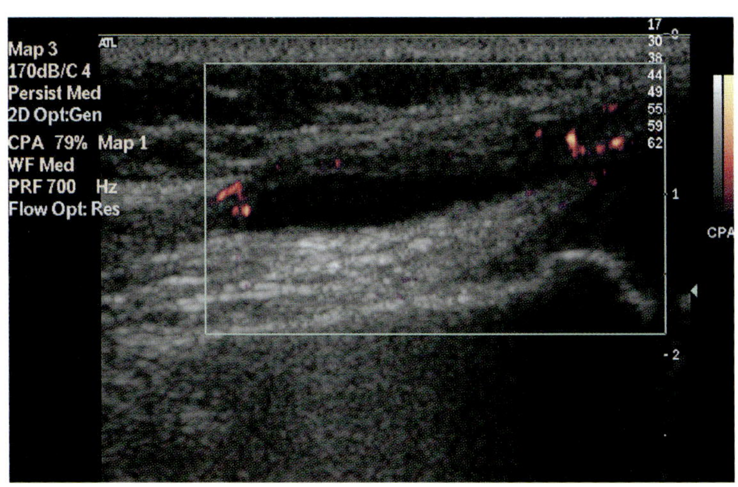

Fig. 4.3 Increased blood flow seen around the suprapatellar pouch in a knee with an effusion and joint infection

haemarthrosis and pus, but also some inflammatory exudates with high protein content. Comparison with the normal side is very useful but care should be taken to place the normal limb in the same position as the suspect one. There is no need to straighten both as an effusion will be best seen in the position that the patient finds most comfortable. Indeed, in the hip straightening forcibly may push the fluid deep into the joint and make it less apparent in the anterior capsule. Position the normal side to match the affected limb.

The most sensitive scanning plane will be along the long axis of the limb over the area where the capsule is most slack. In the hip this is along the anterior femoral neck in an oblique sagittal plane; in the elbow in the posterior joint with the elbow flexed and dependent; in the shoulder along the biceps tendon where the joint is most dependent and loose.

Unfortunately, although US can accurately detect effusions and perhaps more importantly exclude their presence, it cannot determine the nature of the fluid. The echo pattern, the blood supply to the capsule and the size of any effusion may all be assessed, but no one feature is specific for blood or pus as opposed to transudate [7]. This discrimination can only be made by aspirating the fluid.

Patients with Perthes' disease commonly have an effusion but also may show atrophy of the quadriceps muscle when examined by US. This is not a common feature of transient synovitis [8].

4.2.2.4 CT

In children CT is best avoided because of the radiation dose that results. Motion of the patient may be a problem even with modern fast helical machines and sedation may be required. In the detection of joint effusions CT has modest value and should be reserved for those cases where there is bony involvement which must be assessed prior to surgery. Even then it would be in the rarest of cases where a combination of US and MRI were not the preferred approach.

4.2.2.5 MRI

Arguably MR is the most sensitive method for detecting joint effusions but it is impractical for children as a routine method. T2-weighted images show high signal with fluid and capsular distension will be recognized (Fig. 4.4). It can demonstrate the whole joint and not just the areas unmasked by bone. US has blind areas but fortunately this

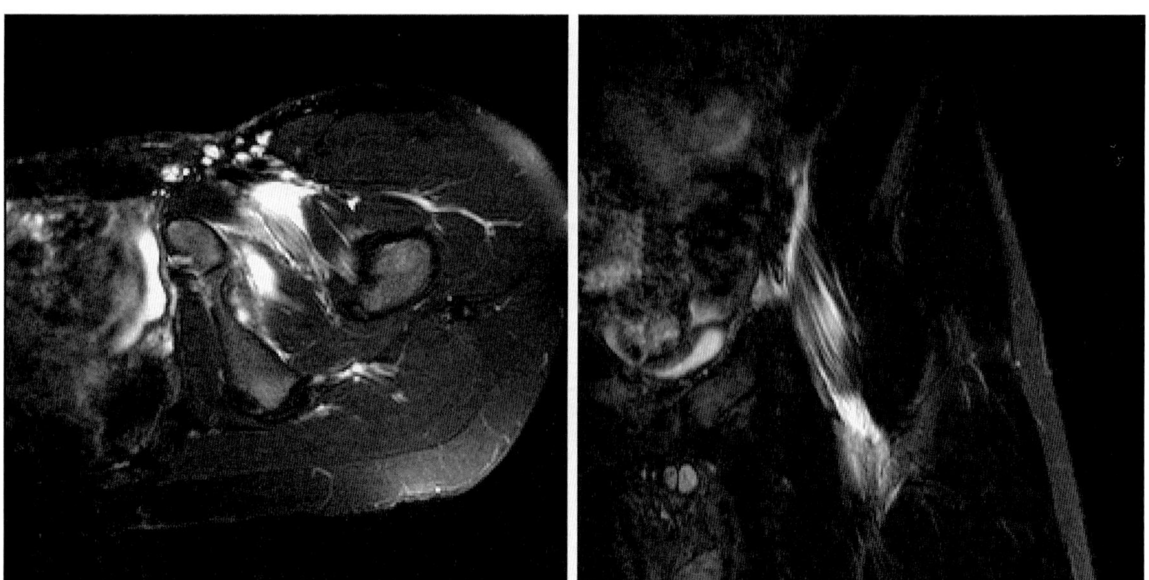
a b

Fig. 4.4a,b Fat-suppressed images of a hip with both a joint effusion and muscle haemorrhage/oedema after trauma

is rarely a practical problem. The major weakness of MR is that it cannot discriminate solid synovial hypertrophy from effusion without using intravenous or intra-articular contrast agents. The manoeuvres are expensive, time consuming and invasive. They should never be used when US is available.

US cannot image osseous disease and the main role of MR is to show osteomyelitis, bone oedema and microfractures. There is evidence that bone oedema does not occur in transient synovitis but may be seen in some but not all cases of septic arthritis [9]. Patients who have symptoms of irritable hip and no effusion on US are best examined by MR. The role of scintigraphy in this context has been completely replaced. MR is more sensitive, more specific and does not require venepuncture or radiation exposure. However, movement of the child may be a problem and carefully supervised sedation or anaesthesia may be necessary.

4.2.3 Role of Imaging in Treatment

Many joints may be aspirated without image guidance. However, the hip and shoulder are deep and US is invaluable in identifying the best point for puncture. Once found it is usually best to place the needle without direct US guidance. This is because US guidance with sterile probe covers takes time and requires gentle manipulation of the needle approach, and children are unlikely to tolerate these manoeuvres. Aspirate the joint to dry as this will alleviate the pain and send material for urgent Gram staining and culture. In selected cases, crystal studies are appropriate. If a joint is fully aspirated the pain relief is far better than can be achieved by analgesia, bed rest or skeletal traction. These measures are now rarely needed. A clear effusion with a negative Gram stain would allow the child to be sent home for outpatient investigation and follow-up, thus saving admission and psychological trauma. The discomfort of a joint aspiration should be little different to a venepuncture, especially if local anaesthetic jelly is applied to the potential puncture site at least 90 minutes in advance of the US examination. To avoid a second visit to the ultrasound room, it is useful to train the ward staff where the puncture site normally is and to ask them to apply the local anaesthetic jelly in advance of the diagnostic procedure [10].

4.2.4 Management Protocol

It pays to have a clear and agreed protocol for the management of a child with an acutely painful joint. Many clinical teams are involved and a clear pathway helps all. Whether the referral is via general practice, accident and emergency, paediatrics or orthopaedics, the process should be the same to avoid doubt, risk and oversight.

The initial assessment is not contentious:

- History from child
- History from parent or guardian
- Clinical examination
- Ultrasound of the affected joint with comparison to the normal side

If neither fluid nor synovitis are found on ultrasound then there should be further investigation to find the source of the symptoms. A plain film should be the next step followed by a clinical review. Examples of disease that may mimic hip pain are iliopsoas abscess, muscle strain or tear, stress fracture and retrocaecal appendicitis, amongst many others.

If the child is over 8 years of age then plain films are mandatory to exclude SUFE and Perthes' disease. Joint effusion is seen in 74% of patients with Perthes' disease and in 50% of those with SUFE. The film should be taken in "frog lateral" position as 11% of cases of SUFE would be overlooked on an ordinary AP (frontal) view.

If there is fluid, then it may be "benign" transudate or "aggressive" sepsis. There is only limited evidence from small numbers and series where there were no infected cases that the severity of the symptoms, ESR, CRP, blood count, body temperature, volume of effusion or the presence of synovial thickening are predictive of whether the fluid is pus or transudate [11]. Fortunately septic arthritis is rare but this makes statistical analysis difficult when asking whether a sign can exclude the condition. Indeed, there are examples of children with small effusions and no synovial reaction who have a normal blood test and are afebrile and who still have a septic arthritis. There remains serious doubt that any combination of US and clinical signs can safely exclude infection [12]. This is the first argument for diagnostic aspiration. Should the effusion be free from infection then the symptoms will resolve after the aspiration as they are due to the pressure inside the joint. Early aspiration has been shown to give a lasting relief of effusion compared to conservative management [13]. This allows early discharge and provides the second argument for early US-guided joint puncture. Counter to this there is a logistics problem in providing what must be a 24-hour US and aspiration service. Some centres elect to admit overnight for observation and analgesia. This carries the risk that the septic joint may deteriorate with risk if irrevocable damage to the articular cartilage. In one year's practice we found that 4% of children with irritable hip were suffering from a septic arthritis (Nuffield Orthopaedic Centre, internal audit data).

Lastly there will be the occasional case where there is a sterile effusion that is drained and then after several days or even weeks that condition recurs [14, 15]. Initially, the same clinical examination and US scheme should be used for the second episode, but it is then wise to arrange an MR study [4, 16] and serology for inflammatory markers. Some cases of recurrent transient synovitis will never be explained but others will be the presentation of juvenile arthropathies or adjacent bone disease with secondary joint irritation. Bone diseases include, osteomyelitis, Perthes' disease and subarticular tumours including osteoid osteoma. Acute chondrolysis is rare but will cause similar symptoms and will only be detected on high resolution MRI, preferably MR arthrography. It is seen as cartilage oedema with focal thickening and later as defects and loss of joint space.

4.2.5 Potential Developments

The increased sensitivity of US to blood flow now leads to the possibility of assessing the amount of blood that is flowing in the periosteal supply to the femoral neck [17]. It may be possible to judge the vascular viability of the femoral head in conditions that threaten its blood supply, notably SUFE, Perthes' disease and fractures. However, the method is not yet reliable as a tool to discriminate septic arthritis from transient synovitis [18]. There are significant problems associated with quantification and it may be that newer microbubble contrast agents may hold the key

Acquisition of 3D data by US may become a method of assessing the shape and size of a joint prior to surgical correction by osteotomy or rotational procedures. Having said this, MR is already more than capable of providing such data and will take some dislodging as the method of choice.

Weight-bearing MR is feasible on purpose-made machines. This should cast better light on the biomechanics of alterations in joint shape and alignment. It may predict those cases liable to early stress-related degenerative disease.

4.3 Synovitis

4.3.1 Clinical Background

The above description may be seen as a model for the investigation and management of children with an isolated arthropathy. There will be occasions where multiple joints are involved. The differential diagnosis includes juvenile arthropathy due to a collagen disorder, reactive arthropathy secondary to an infection (gastrointestinal in particular) and rarely coagulation disorders. The diagnosis will be made on the basis of a spectrum of findings from history, clinical examination, serology and imaging. Of these, imaging is perhaps the least effective. However, there is a much stronger role for the radiologist who can determine the severity of the synovial disease, establish its spread or extent and demonstrate changes over a period of time.

4.3.2 Role of Imaging

The questions posed of imaging are:

- Is there synovitis?
- Which joints are affected?
- Is there bone or cartilage damage?
- How is the disease progressing?

Plain films will demonstrate the destructive marginal changes of established aggressive synovial disease but they are usually only positive at a stage when the opportunity to treat effectively with disease-limiting drugs has passed or at least reached a point of considerably diminished return. In children, marginal erosions will affect unossified bone and will be very late to present.

Ultrasound is the undoubted method of choice in detecting joint effusions. It also is the best way of differentiating joint fluid form synovial thickening [19]. These look almost the same on unenhanced MR images. US with colour flow Doppler, preferably power Doppler, will show the extent of the vascular supply of the synovium providing a qualitative representation of the degree of synovial inflammation. US is particularly effective in mapping the number and distribution of joints involved and has proved to be better than plain films and clinical examination in grading the involvement of joints [20] with a sensitivity seven times that of the older methods [21]. The experienced examiner can achieve a rapid tabulation of involvement of all the joints of the extremities in less than 20 minutes.

MRI can only be used to differentiate fluid form synovial hypertrophy if intravenous Gd-DTPA or intra-articular injection is performed, but it may well be the most sensitive method that we have to do this [22] (Fig. 4.5). If the question is whether synovitis is present then conventional unenhanced FSTIR images may show erosion, avoiding the need for injection. However, MRI is less useful for plotting the extent of disease as, although it would work, it would need multiple images after contrast and would take an impractical length of time in patients whose patience and cooperation will deteriorate rapidly.

4.3.3 Potential Developments

The roles of US and MR in following the progress of drug treatment have not been fully established but there is likely to be an important role both in systemic treatment and intra-articular injection [23]. There are problems in quantifying the degree of vascular injection using both techniques. Temperature changes in the room, the position of the patient and pressure exerted by the imaging devices will all alter the blood supply. US contrast agents may give a more reproducible method and MR Gd-DTPA uptake measurements have potential [24].

4.4 Osteomyelitis and Soft-tissue Infection

4.4.1 Clinical Background

Bone infections may result from direct penetration and implantation in the bone from trauma or surgical procedures. Less common but potentially serious bone infection of haematogenous origin presents with pain, dysfunction and sometimes fever. Whatever the origin, bone infection is difficult to eradicate and chronic infection often results. A careful diagnostic work-up, staging by imaging, microbio-

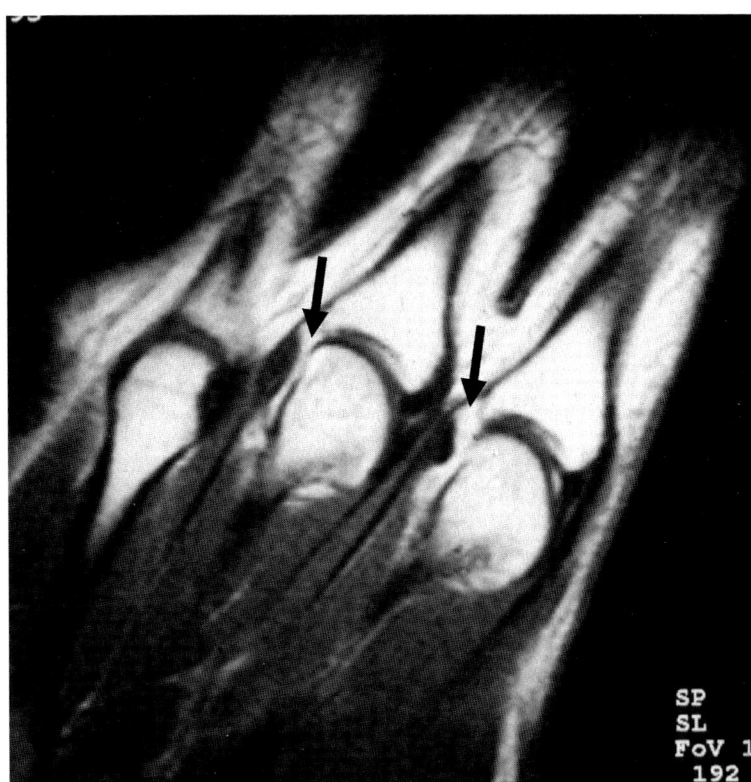

Fig. 4.5 T1-weighted MR after intravenous Gd DTPA in early synovitis (*arrows* white area) with normal plain films. Compare the adjacent joints

logical typing and tailored antibiotic and surgical treatment are essential. All too often bone infection is poorly or incompletely treated, leading to chronic and often very long-term disease.

In those cases where there is direct implantation, the organism will depend on the degree and type of contamination. Gram-positive cocci are common but many types of infection are seen. Those infections of hospital origin may well be resistant to a number of antibiotics. If the wound is poorly vascularized or frankly necrotic, different organism abound. Gram-negative rods and even gas-forming organisms such as clostridia may be found. An unusual infection, found in those who keep fish, is tuberculosis marium, found in fish tanks and caught by handling the contents with a wound or cut on the hand.

Infections of blood-borne origin are typically due to *Staphylococcus aureus*, *Streptococcus pyogenes* and *Mycobacterium tuberculosis*. It can be argued that tubercular infections tend to be lower grade clinically with less fever and a subacute presentation. However, there is so much overlap that it is wise to assume that any individual case of bone infection could be due to pyogenic and tubercular infection whatever the history and signs at presentation.

Chronic recurrent multifocal osteomyelitis in children will mimic acute and chronic osteomyelitis clinically and on imaging [25]. The lack of an identifiable organism, resolution and recurrence at other sites are later but diagnostic features. There is an association with pustular skin diseases. Imaging by plain films and MRI will define the extent but will not allow a diagnosis. Bone scintigraphy is arguably the easiest way of searching for other occult or early sites; all the long bones should be included in the examination.

Soft-tissue infections range from cellulitis to abscess formation. Again the pyogenic organisms predominate. There is a strong link with penetrating injuries. Muscle infection may be secondary to diffuse soft-tissue cellulitis or to osteomyelitis. It may also occur spontaneously without other compartments involved, especially in immunosuppressed patients.

4.4.2 Role of Imaging in Detection

Plain films are sometimes diagnostic with evidence of periosteal reaction and permeative lysis in the affected areas. Later, repair and necrosis may lead to sclerosis, new bone formation and even frank destruction. Unfortunately, in the early stages the images are normal and it may take several weeks for plain films to reveal any signs.

Isotope bone scintigraphy has been used extensively to detect areas of increased bone turn-over suggesting that infection is a cause. The changes are nonspecific and there is little to differentiate the signs resulting from trauma, infection, tumour and degeneration. The greatest value of scintigraphy is to declare that the bones are normal having understood that there may be false-negative examinations in the first few days of osteomyelitis. Gallium citrate scintigraphy and indium chloride-labelled white cell studies are more specific for infection but there remains sufficient overlap with other diseases to require a biopsy for confirmation in most cases. Therefore as a biopsy is likely in any case where the imaging is focally abnormal in an area that cannot be explained by other factors then there is little value in tests that merely increase the chance that infection is the cause. The biopsy will do this whatever the imaging shows.

Ultrasound has a relatively minor role in the staging and confirmation of osteomyelitis. However, it may detect periosteal reactions as a thin layer of fluid alongside the periosteum even in cases where the plain film is abnormal (Fig. 4.6). This sign should lead to further investigation, typically plain films and MRI, usually followed by a biopsy. The converse significance of this sign is not helpful as there are cases of proven osteomyelitis where there was no periosteal collection seen with US. US is the technique of choice in differentiating cellulitis from abscess cavitation and in determining whether the joints are involved [26]. US is very sensitive and accurate in detecting pyomyositis but MR may be needed to exclude osteomyelitis [27].

Bone scintigraphy used to be the investigation chosen to rule out osteomyelitis when plain films were normal but it has been supplanted in this by MRI. There may be rare occasions when MR is impractical or equivocal when triple-phase bone scintigraphy has a role [28].

MRI is the most useful investigation for staging the extent of disease [29–31] (Fig. 4.7). A normal study using T1-weighted and T2-weighted images with fat suppression (or STIR) will exclude even early symptomatic osteomyelitis [32]. The only occasion where infection is likely to be overlooked is when the disease is not yet producing symptoms. Areas of oedema (low on T1 and high on T2) are also seen in microfractures due to trauma. Infection typically leads to adjacent soft-tissue changes but as these are common in injuries, this does not help. The history

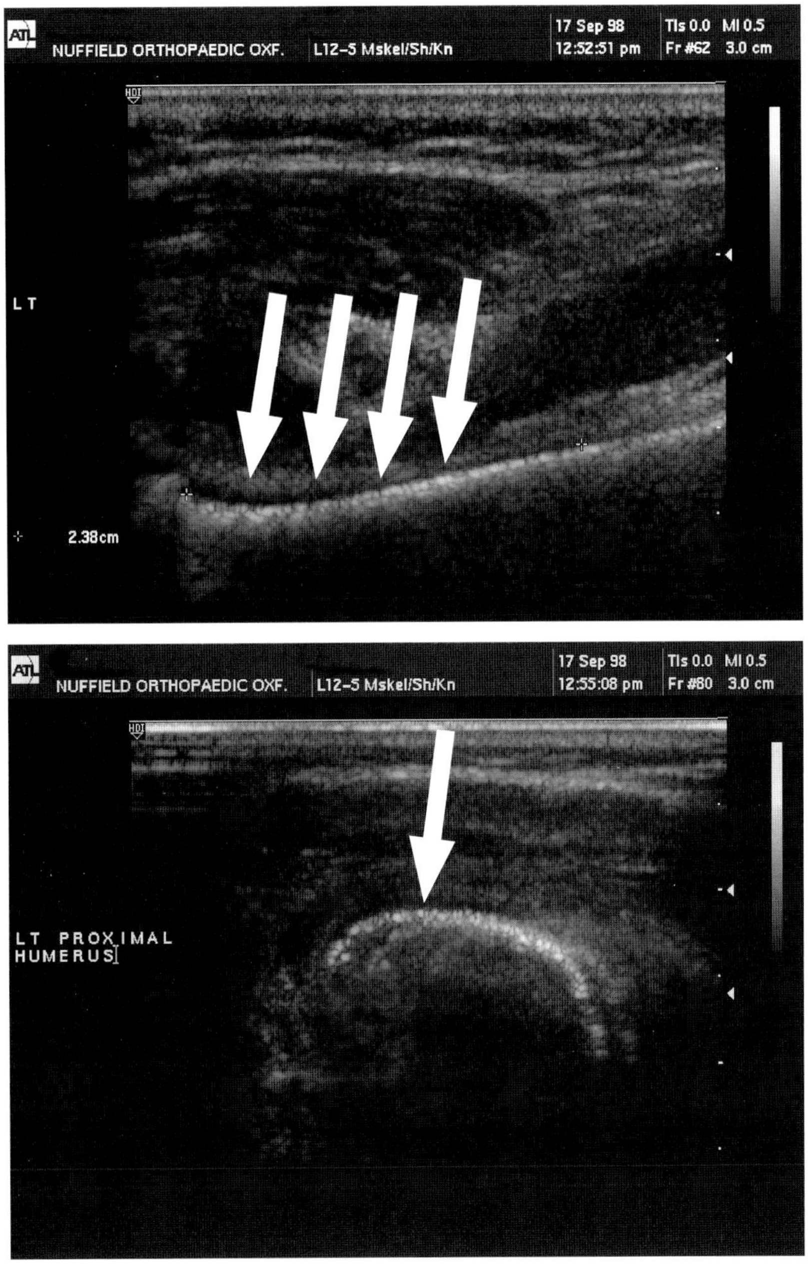

Fig. 4.6a,b Sagittal and axial US showing a periosteal reaction due to proximal femoral acute osteomyelitis. Plain films were normal at this stage

is critical but this may not always be reliable in children or for that matter in adults. The extent of infection including involvement of joints, abscess formation and necrosis of the affected bone are all seen on MR (Figs. 4.8, 4.9). The areas infected tend to be surrounded by oedema and it is not possible to differentiate this from truly infected regions. Intravenous contrast may increase the ease of discrimination of oedema from infected areas but it is not reliable and rarely makes any impact on management decisions. Biopsies should be taken bearing in mind that the margins may not be representative whilst the centre of the lesion may be necrotic. It is prudent to take several samples from different areas. MRI is poor at discriminating oedema from a fluid-containing abscess; intravenous contrast helps but may be confusing. Ultrasound is the best method of showing pus or fluid that could be aspirated or removed and is the ideal means of guiding needle puncture or catheter introduction.

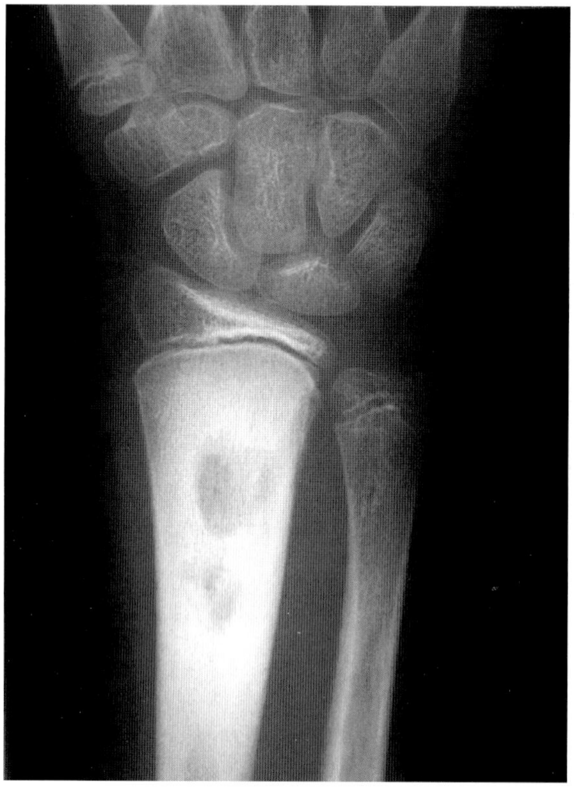

a

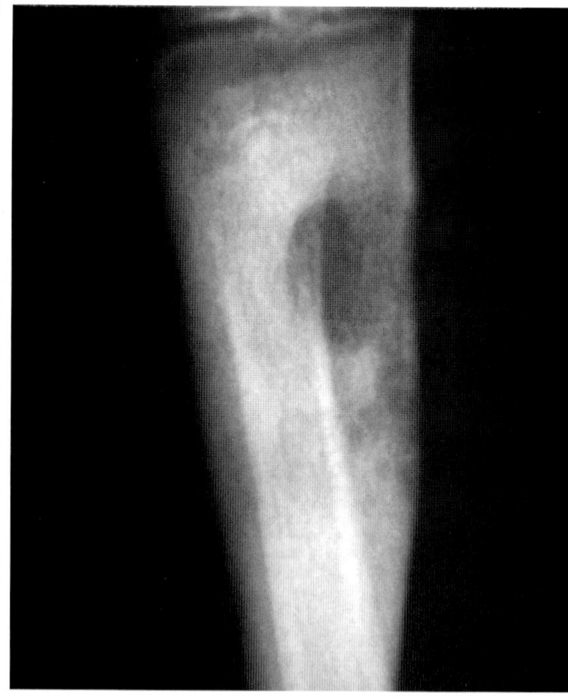

b

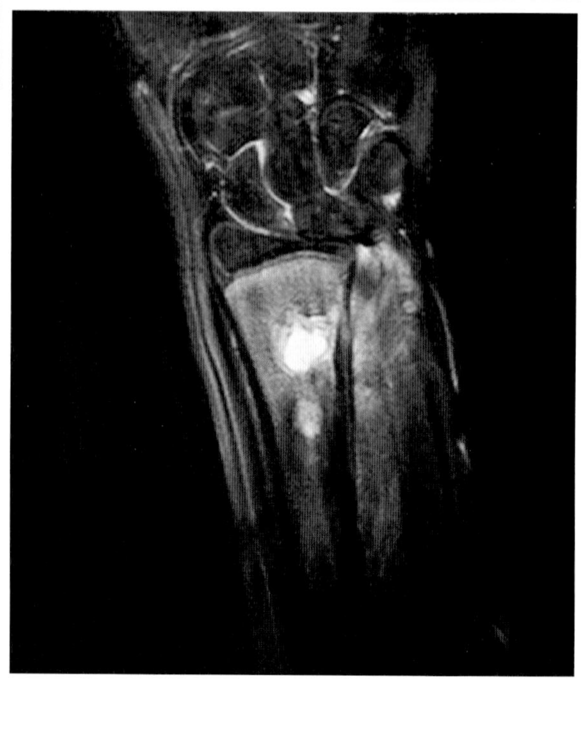

c

Fig. 4.7a–c Plain films and MR (FSTIR) of a case of Brodie's abscess (chronic osteomyelitis with sclerosis around an intraosseous abscess cavity)

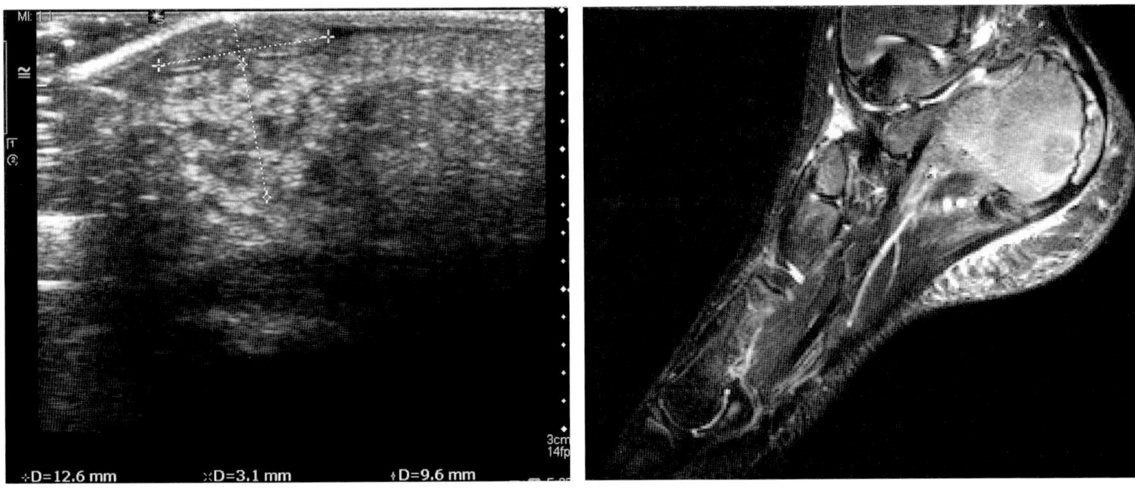

a b

Fig. 4.8a,b US and MR of a patient with a small abscess containing echogenic (white) pus adjacent to the calcaneus. FSTIR MRI shows oedema in the adjacent bone which could be reactive or a sign of early osteomyelitis. Follow-up MR after antibiotic treatment showed resolution of this issue

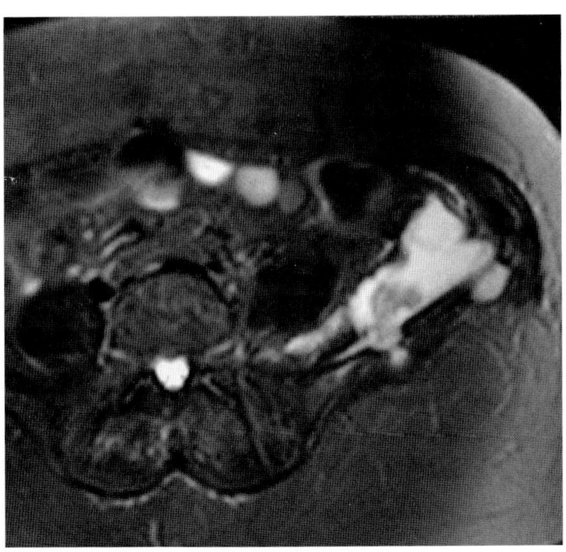

a

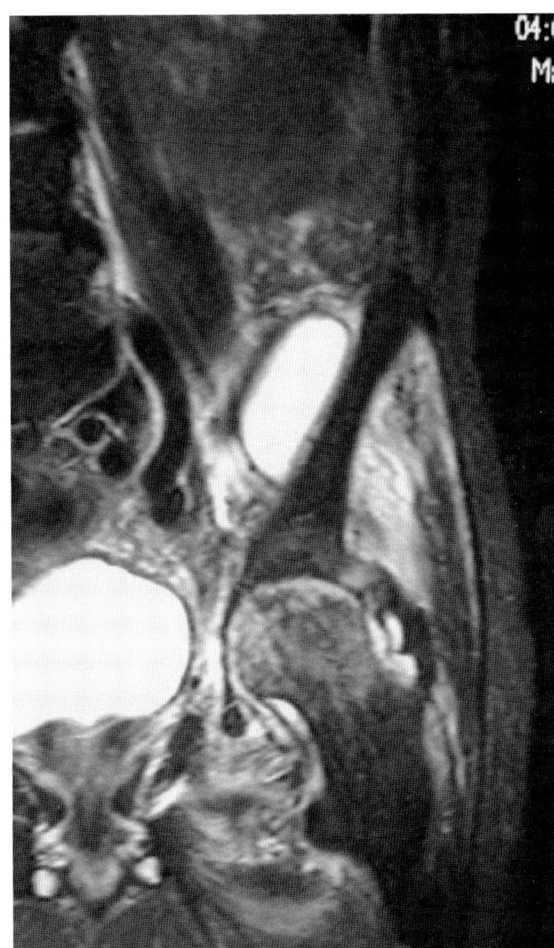

b

Fig. 4.9 Tuberculous abscess on the medial side of the ileum and pyomyositis of the proximal gluteal muscles on the lateral side of the bone that is also infected

4.4.3 Role of Imaging in Treatment

As detailed above, imaging, especially MRI, is important in deciding where to biopsy. US will determine if the mass effect seen on MR or detected clinically is an abscess. Most will use fluoroscopy or CT guidance to obtain the specimens and US to aspirate fluid. The choice depends on local facilities and individual skills, but fluoroscopy is generally faster and less complex for bone biopsy and US is more suitable for soft-tissue lesions. CT is essential when there are intervening gas-filled structures that must be avoided.

MR is the best means of following the response to treatment; it is best performed on the same equipment with the same protocols for each follow-up study. Even with extended field of view imaging, US is poor at providing images that can be compared in follow-up examinations. It may be used to review changes in overall dimensions of mass lesions or cavities. Surgical planning will be based on a combination of imaging methods, but MR is the principle method [33].

4.4.4 Potential Developments

Advances in MR techniques including diffusion-weighted imaging and chemical shift imaging may increase the precision in deciding which area is truly infected and which is only reactive oedema. This may influence borderline decisions to operate, especially when the area of surgery must be limited because of adjacent critical structures.

References and Further Reading

1. Eich GF, Superti-Furga A, Umbricht FS, et al (1999) The painful hip: evaluation of criteria for clinical decision-making. Eur J Pediatr 158(11):923–928
2. Kemp HS, Boldero JL (1966) Radiological changes in Perthes' disease. Br J Radiol 39(466):744–760
3. Castriota-Scanderbeg A, Orsi E, De Micheli V, et al (1993) Ultrasonography in the diagnosis and follow-up of hip pain in children (in Italian). Radiol Med (Torino) 86(6):808–814
4. Shih TT, Su CT, Chiu LC, et al (1993) Evaluation of hip disorders by radiography, radionuclide scanning and magnetic resonance imaging. J Formos Med Assoc 92(8):737–744
5. Wilson DJ, Green DJ, MacLarnon JC (1984) Arthrosonography of the painful hip. Clin Radiol 35(1):17–19
6. Adam R, Hendry G, Moss J (1986) Arthrosonography of the irritable hip in childhood: a review of 1 years experience. Br J Radiol 59:205–208
7. Konermann W, Gruber G (1997) [Septic coxitis in childhood. Differential ultrasound diagnoses]. Orthopade 26(10):830–837
8. Robben SG, Meradji M, Diepstraten AF, et al (1998) US of the painful hip in childhood: diagnostic value of cartilage thickening and muscle atrophy in the detection of Perthes' disease. Radiology 208(1):35–42
9. Lee SK, Suh KJ, Kim YW, et al (1999) Septic arthritis versus transient synovitis at MR imaging: preliminary assessment with signal intensity alterations in bone marrow. Radiology 211(2):459–465
10. Berman L, Fink AM, Wilson D, et al (1995) Technical note: identifying and aspirating hip effusions. Br J Radiol 68:306–310
11. Skinner J, Glancy S, Beattie TF, et al (2002) Transient synovitis: is there a need to aspirate hip joint effusions? Eur J Emerg Med 9(1):15–18
12. Luhmann SJ, Jones A, Schootman M, et al (2004) Differentiation between septic arthritis and transient synovitis of the hip in children with clinical prediction algorithms. J Bone Joint Surg Am 86(5):956–962
13. Kesteris U, Wingstrand H, Forsberg L, et al (1996) The effect of arthrocentesis in transient synovitis of the hip in the child: a longitudinal sonographic study. J Pediatr Orthop 16(1):24–29
14. Ranner G, Ebner F, Fotter R, et al (1989) Magnetic resonance imaging in children with acute hip pain. Pediatr Radiol 20(1–2):67–71
15. Terjesen T, Osthus P (1991) Ultrasound in the diagnosis and follow-up of transient synovitis of the hip. J Pediatr Orthop 11(5):608–613
16. Minio Paluello GB, De Pellegrin M, Tacchini S, et al (1995) Persistent coxalgia in the child. The value of magnetic resonance (in Italian). Radiol Med (Torino) 89(4):402–408
17. Robben SG, Lequin MH, Diepstraten AF, et al (2000) Doppler sonography of the anterior ascending cervical arteries of the hip: evaluation of healthy and painful hips in children. AJR Am J Roentgenol 174(6):1629–1634
18. Strouse PJ, DiPietro MA, Adler RS (1998) Pediatric hip effusions: evaluation with power Doppler sonography. Radiology 206(3):731–735
19. Wakefield RJ, Kong KO, Conaghan PG, et al (2003) The role of ultrasonography and magnetic resonance imaging in early rheumatoid arthritis. Clin Exp Rheumatol 21(5 Suppl 31):S42–S49
20. Szkudlarek M, Narvestad E, Klarlund M, et al (2004) Ultrasonography of the metatarsophalangeal joints in rheumatoid arthritis: comparison with magnetic resonance imaging, conventional radiography, and clinical examination. Arthritis Rheum 50(7):2103–2112
21. Kane D, Grassi W, Sturrock R, et al (2004) Musculoskeletal ultrasound—a state of the art review in rheumatology. Part 2: Clinical indications for musculoskeletal ultrasound in rheumatology. Rheumatology (Oxford) 43(7):829–838
22. Hoving JL, Buchbinder R, Hall S, et al (2004) A comparison of magnetic resonance imaging, sonography, and radiography of the hand in patients with early rheumatoid arthritis. J Rheumatol 31(4):663–675
23. Filippucci E, Farina A, Carotti M, et al (2004) Grey scale and power Doppler sonographic changes induced by intra-

articular steroid injection treatment. Ann Rheum Dis 63(6):740–743

24. Salaffi F, Carotti M, Manganelli P, et al (2004) Contrast-enhanced power Doppler sonography of knee synovitis in rheumatoid arthritis: assessment of therapeutic response. Clin Rheumatol 23(4):285–290
25. Schilling F, Kessler S (2001) Chronic recurrent multifocal osteomyelitis—I. Review (in German). Klin Padiatr 213(5):271–276
26. Robben SG (2004) Ultrasonography of musculoskeletal infections in children. Eur Radiol Jan 30 (Epub ahead of print)
27. Trusen A, Beissert M, Schultz G, et al (2003) Ultrasound and MRI features of pyomyositis in children. Eur Radiol 13(5):1050–1055
28. Connolly LP, Connolly SA, Drubach LA, et al (2002) Acute hematogenous osteomyelitis of children: assessment of skeletal scintigraphy-based diagnosis in the era of MRI. J Nucl Med 43(10):1310–1316
29. Teo HE, Peh WC (2004) Skeletal tuberculosis in children. Pediatr Radiol 34:853–860
30. Marin C, Sanchez-Alegre ML, Gallego C, et al (2004) Magnetic resonance imaging of osteoarticular infections in children. Curr Probl Diagn Radiol 33(2):43–59
31. Kleinman PK (2002) A regional approach to osteomyelitis of the lower extremities in children. Radiol Clin North Am 40(5):1033–1059
32. Aloui N, Nessib N, Jalel C, et al (2004) Acute osteomyelitis in children: early MRI diagnosis (in French). J Radiol 85(4 Pt 1):403–408
33. Santiago Restrepo C, Gimenez CR, McCarthy K (2003) Imaging of osteomyelitis and musculoskeletal soft tissue infections: current concepts. Rheum Dis Clin North Am 29(1):89–109

5 Soft Tissue Tumours in Children

Gina Allen

CONTENTS

5.1 Introduction 67
5.2 US 67
5.2.1 Plain Radiographs 69
5.2.2 Magnetic Resonance Imaging 69
5.3 Computed Tomography 70
5.4 Nuclear Medicine 70
5.5 Disease characteristics 71
5.5.1 Benign Lesions Seen On US 71
5.5.1.1 Lymph Nodes 71
5.5.1.2 Ganglia 73
5.5.1.3 Popliteal Cysts 73
5.5.1.4 Lipomas 74
5.5.1.5 Sebaceous Cysts 75
5.5.1.6 Verrucas 75
5.5.1.7 Foreign Bodies 75
5.5.2 Vascular Anomalies 76
5.5.2.1 Infection 78
5.5.2.2 Muscle Hernias 79
5.5.2.3 Malignant Lesions 79
5.6 Potential Developments 82
References and Further Reading 82

5.1 Introduction

The finding of a lump by a child or parent is always met with anxiety. Although a good history and examination can often go a long way in determining whether a lesion needs further investigation, imaging can often completely reassure the patient and parents that the lesion is benign, permitting the family to go away "happy". Whilst soft tissue tumours in children are rarely malignant, it is important to make an early diagnosis when they are!

G. Allen, DCH, MRCP, FRCR
The Royal Orthopaedic Hospital NHS Trust, Bristol Road South, Northfield, Birmingham, B31 2AP, UK

5.2 US

Ultrasound (US) examination should be the first-line imaging investigation after a soft tissue "lump" has been discovered. It does not involve radiation or sedation. The precise area of concern can be "pointed to" by the patient and identified prior to imaging. The family is given immediate feedback and unfounded anxieties may be eased.

US examination can detect whether the swelling is purely cystic or has solid elements. Sometimes the location of a mass and its composition can be diagnostic; for example a popliteal cyst – a purely cystic lesion behind the knee arising from between the medial head of the gastrocnemius muscle and the semimembranosus tendon. Similarly a ganglion around the wrist may be safely diagnosed using US alone [1].

In general any lesion that is solid (echogenic) or partly solid should be investigated further. A possible exception is in the case of a haematoma. When there is a clear history of injury and a partly echogenic lesion is found, then it often may be fully assessed by US. However, this must only be performed with considerable caution as there are occasions where a malignant lesion haemorrhages, hiding the original neoplasm. As a consequence it is hazardous to make a definite diagnosis of haematoma from a single US examination. A haematoma, in the early phases, can be solid and echogenic but it will liquefy over the next few weeks and then resolve. This inevitable alteration in pattern can be used to diagnostic advantage. Following such a lesion by additional US examination can be invaluable—watching the "swelling" liquefy over a period of days or weeks until its resolution (Fig. 5.1). The clinical history is also an aid but again not definitive. An area of injury that becomes "softer" and "smaller" is in keeping with a haematoma. A lump should never be labelled as a haematoma without confirmation of resolution by serial US examinations, thereby avoiding the rare but significant risk of overlooking a malignant tumour with secondary

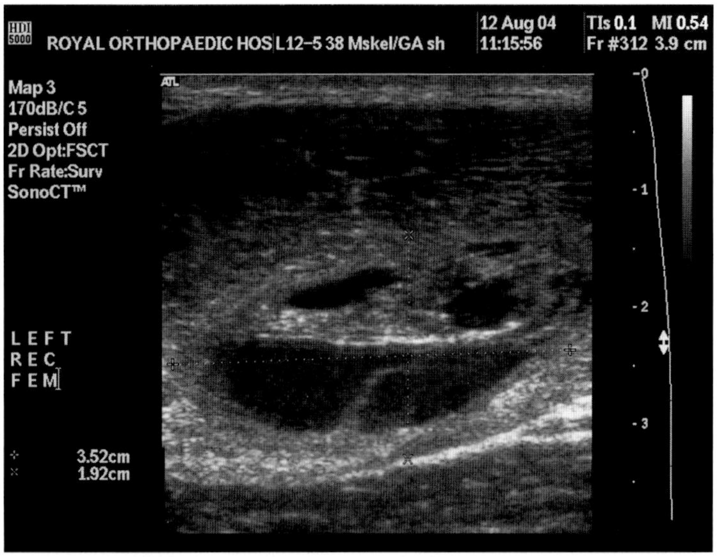

a

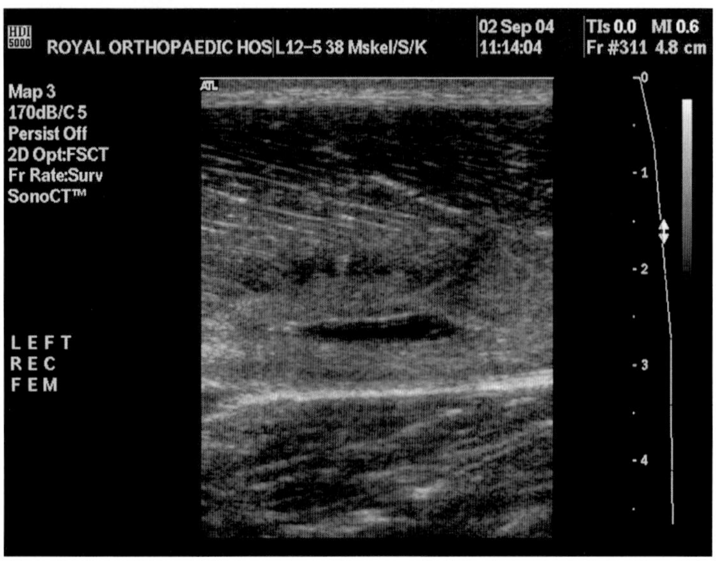

b

Fig. 5.1 a A possible haematoma seen on US as a mixed solid and cystic lesion. **b** Three weeks later the lesion has decreased in size and has completely liquefied, confirming the diagnosis of a haematoma

haemorrhage. In addition, a lesion that is getting larger and "harder" is worrying whatever the imaging shows. When resolution is not steady and progressive it is prudent to perform MR and consider biopsy.

On some occasions the US study prompts further imaging. For example, if the adjacent bone is involved or if the "lump" is uniformly solid in nature and not likely to be a haematoma.

Lipomas are common and usually benign. It would be unreasonable to perform biopsies on all such lesions. A practical and safe approach is to use US to define the nature, size and spread. If the lesion is not growing, lies within a normal fatty layer, does not invade muscle or other non-fat structure and it is less than 10 cm in maximum diameter, then a diagnosis of benign lipoma may be safely made using a combination of clinical and US features. Any variation should be investigated with MR and biopsy. Imaging may help guide the biopsy as heterogeneous tissue may contain areas that are more or less aggressive. Features that suggest that the region is more aggressive and therefore the best site for biopsy are a high interstitial water content (where there is reduction in echogenicity), focal enlargement and invasion of normal structures. "Cystic" areas are to be avoided as these are likely to contain necrotic tissue and will not render diagnostic material.

US can also detect calcification in its very early stages often before it is visible on plain films. It may therefore be useful in some diagnoses such as phleboliths occurring in a haemangioma and calcification in early myositis ossificans (Fig. 5.2).

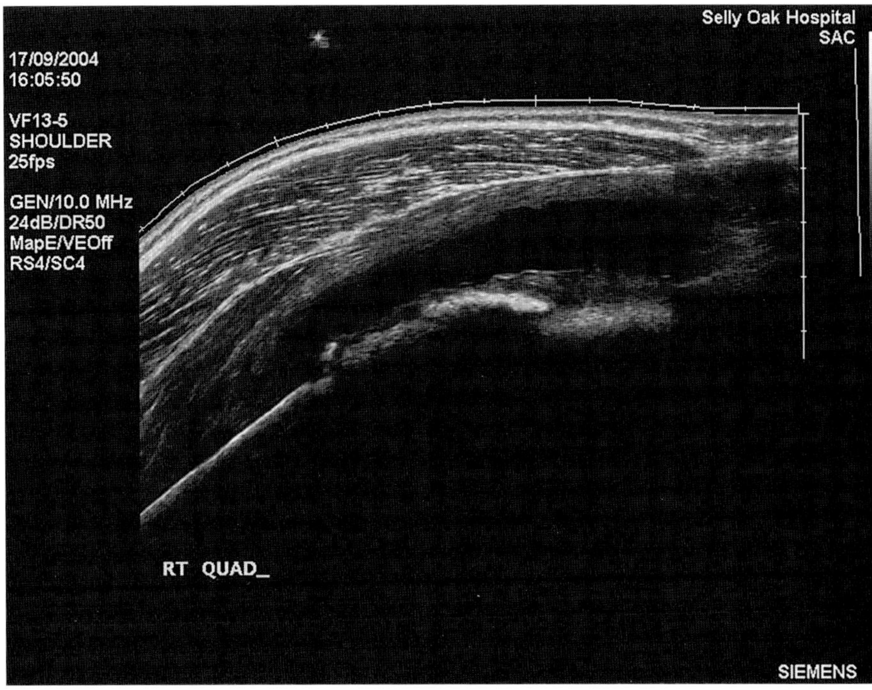

Fig. 5.2 Extended field of view US of a liquefied haematoma with adjacent calcification in keeping with myositis ossificans

5.2.1
Plain Radiographs

Conventional radiographic examination is important in the assessment of a soft tissue lesion, especially if the patient also has bony pain or if the mass appears adherent to the bone on examination. It can also be useful if the lesion appears hard to touch and shows some calcification on US. Soft tissue lesions can be present in primary bone tumours such as a Ewing's sarcoma or infection. Here the site and appearance of the plain films can be diagnostic. Some soft tissue lesions contain calcium deposits which may also be seen on the plain film (for example, phleboliths and myositis ossificans as discussed above) (Fig. 5.3). The detection of calcium in soft tissues may also help in short-listing potential diagnoses. Early calcification will be seen on US some time before it casts a radiological shadow.

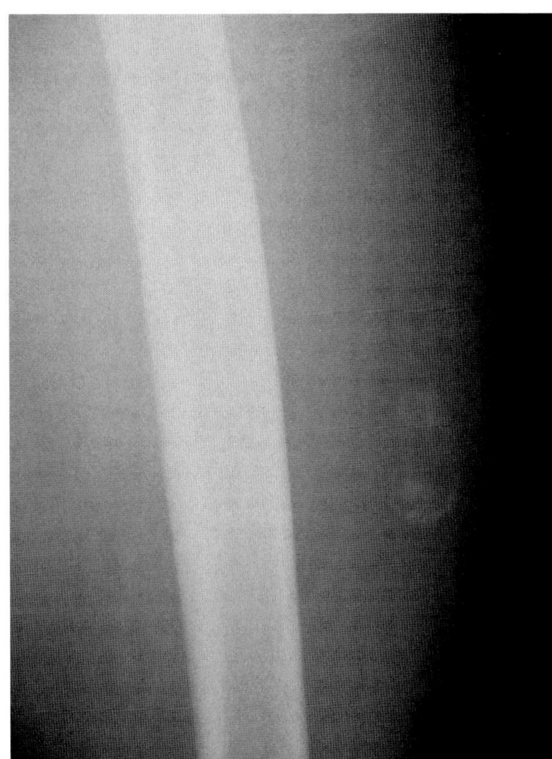

Fig. 5.3 Plain radiograph of myositis ossificans

5.2.2
Magnetic Resonance Imaging

MR examination is essential in locally staging a soft tissue lesion prior to biopsy. It will answer the following questions: Is there subtle involvement of bone? Is this a primary soft tissue or bone lesion? Are there further lesions in the same limb? (skip metastasis) (Fig. 5.4). Signal characteristics often suggest the composition of the mass, whether it is fluid, fibrous, calcified or contains fat. The main difficulty with MRI in children is that it may be necessary to sedate the patient. For this reason this examination should be performed in a specialist unit where the radiographers are experienced in examining children and where monitoring with paediatric anaesthetic cover is available. Contraindications to MR are less likely in children, but some cardiac and intracranial shunts may be a problem. Luckily claustrophobia is less of a problem in children and the majority of children will undergo MR imaging without sedation.

5.3 Computed Tomography

If the lesion arises from bone, CT is sometimes performed, but in children its use should be discouraged as the radiation dose is high. MR will almost always define any bone invasion as well as, if not better than, CT. CT is principally used for bone-based lesions such as osteoid osteoma (Fig. 5.5). CT is sometimes used to measure the limb for custom-made prosthetic replacement of bone, prior to resection of bone tumours. Once the diagnosis of malignancy is made, body CT (chest and sometimes abdomen) is important in the staging of sarcomas. The presence or absence of lung metastasis must be known before embarking on therapy, and therefore a CT of the chest is indicated after histological confirmation of this diagnosis.

Rarely CT of the primary lesion may be the only option, for example when MRI is contraindicated.

5.4 Nuclear Medicine

Bone scintigraphy for metastasis screening is performed to ensure that bone-based lesions are solitary. It may also help detect skip metastasis within the same bone (Fig. 5.6). The presence of multiple lesions suggests metastasis although infection and multiple benign lesions may mimic secondary deposits. Scintigraphy is time consuming and uses radiation and venepuncture. There may be an argument for whole-body MR as an alternative screening tool as MR technology improves and the technique becomes more available [2].

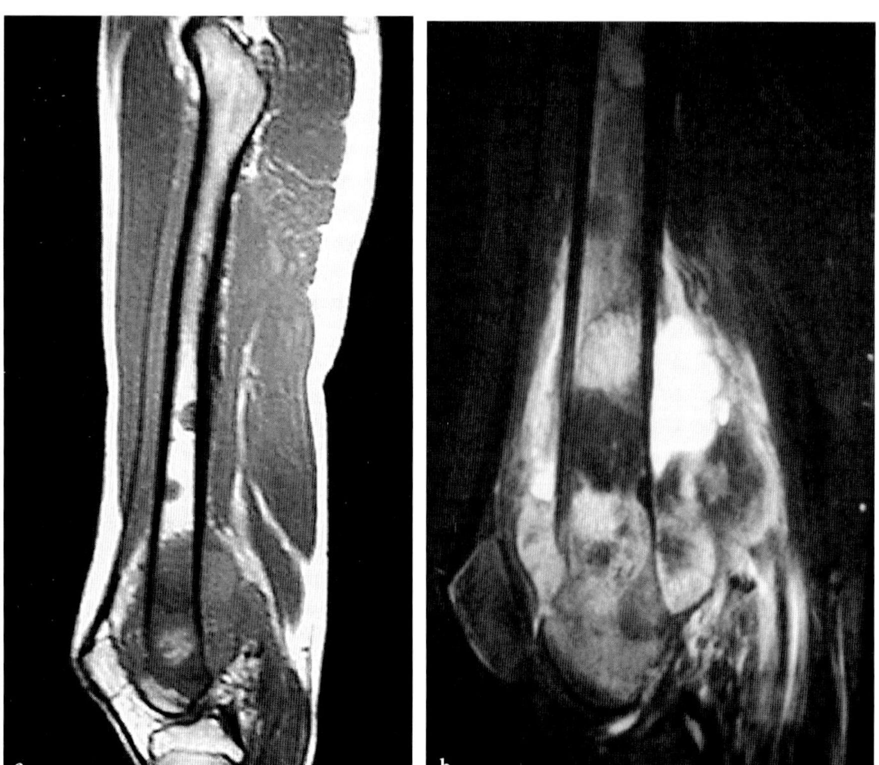

Fig. 5.4a,b MRI showing a primary Ewing's sarcoma with a skip metastasis more proximally

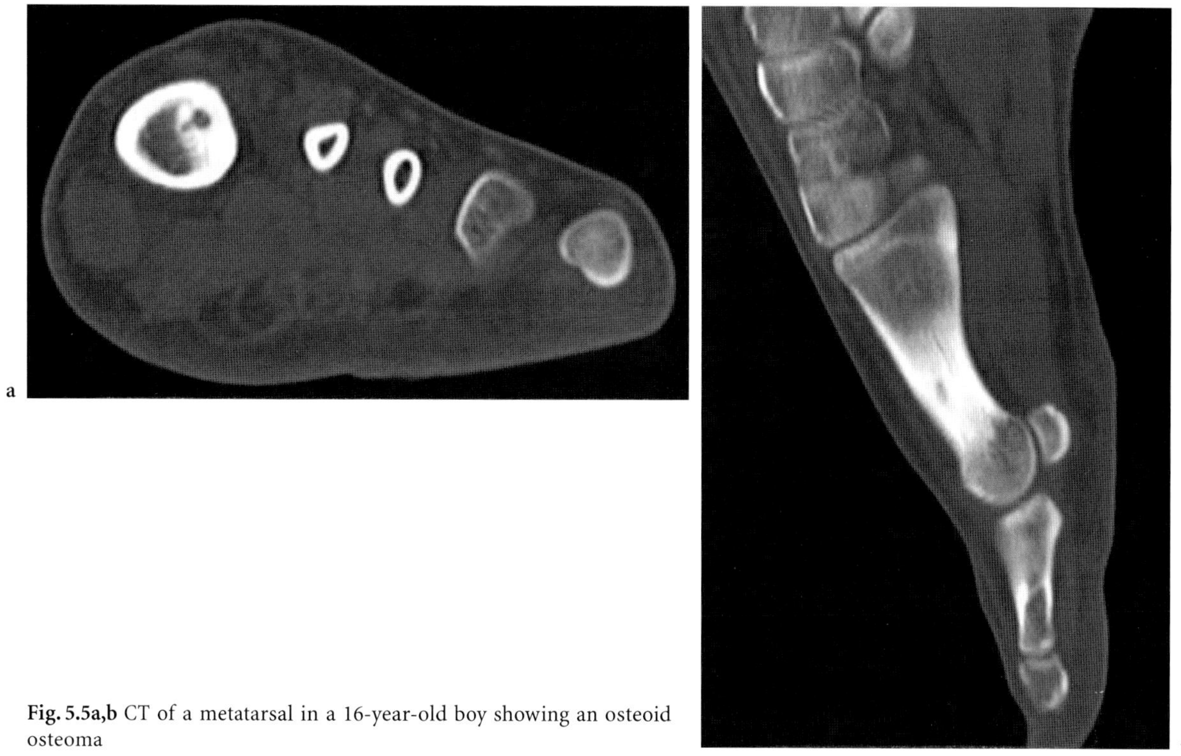

Fig. 5.5a,b CT of a metatarsal in a 16-year-old boy showing an osteoid osteoma

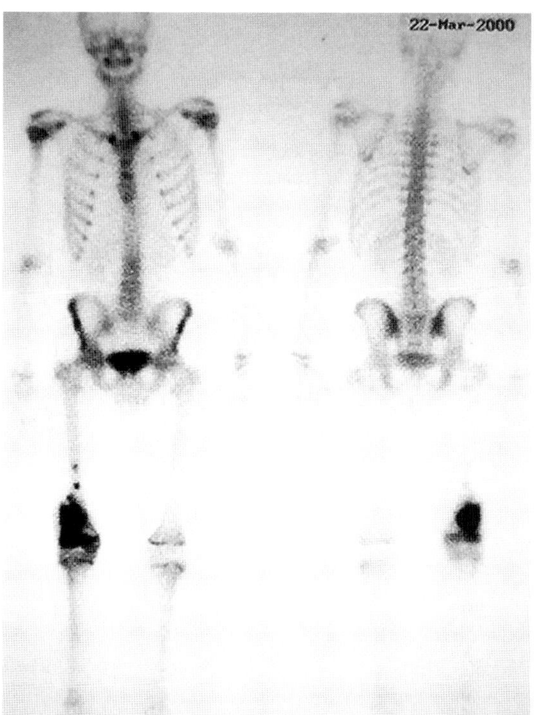

Fig. 5.6 Nuclear medicine Tc99m bone scan showing a single distal femoral lesion and skip metastasis more proximally

5.5 Disease characteristics

5.5.1 Benign Lesions Seen On US

5.5.1.1 Lymph Nodes

Lymph nodes may be found in a variety of locations in children especially in the cervical region. Numerous nodes can be present but if they are in a chain then this is reassuring.

On US, lymph nodes are well-defined, homogeneous lesions usually near neurovascular bundles. They exhibit a characteristic vascular pattern with a large amount of vascularity centrally (Fig. 5.7) [3, 4], and if they are benign and reactive in nature they often have a highly echogenic hilum (Fig. 5.8) [5]. Benign nodes are small and are wider than they are deep. Malignant lymph nodes are rounded, large, of low echogenicity and rarely have an echogenic hilum. They are more likely to have identifiable peripheral vascularity on colour Doppler.

The location of the lesion can be diagnostic. If there is a primary lesion in the area that drains to

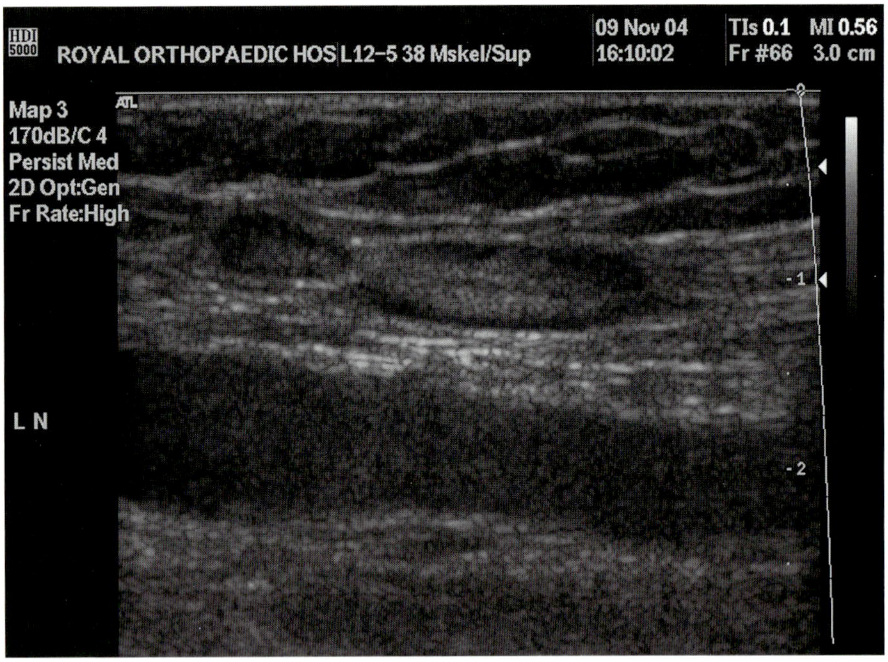

Fig. 5.7 Normal lymph node seen with US with an echogenic hilum

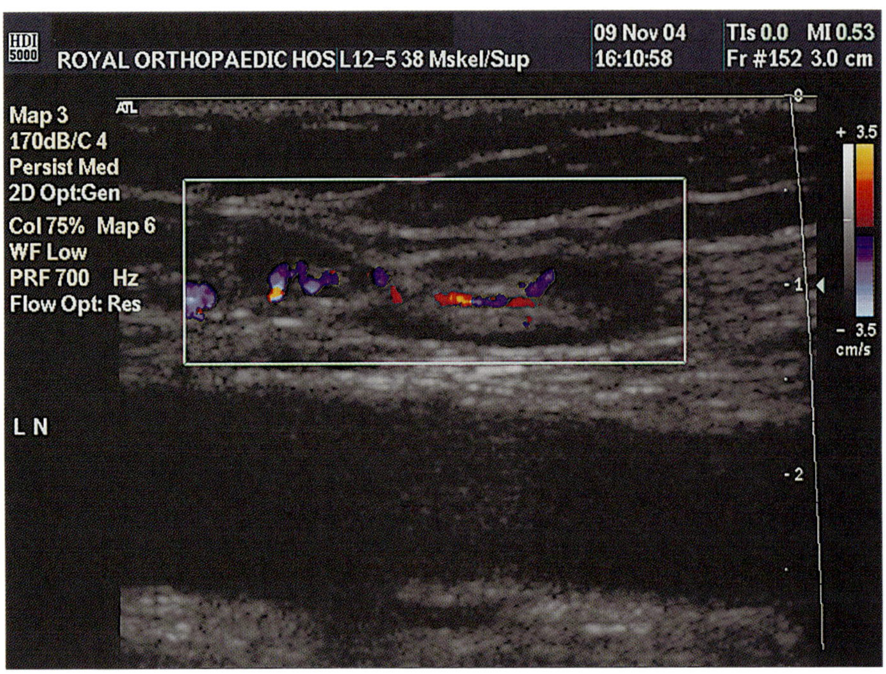

Fig. 5.8 Normal lymph node seen with US and showing central vascularity on colour Doppler

a lymph node group then the clinical detection may explain all the signs without any imaging being necessary. For example, lymph nodes in the groin are usually the result of disease of the lower leg and pelvis. Lymph nodes in the axilla drain the upper limbs and upper trunk. Cervical lymph nodes are associated with primary disease of the head and neck region.

In children most lymph node enlargements, especially widespread lesions, are the result of infection and therefore signs or symptoms of fever, rashes and arthralgia should be sought. They can be present in all the areas mentioned above. Infectious mononucleosis remains a common condition of young people and can cause significant cervical lymphadenopathy. Other important disease processes that cause lymph node enlargement are arthritis (Still's disease) and lymphomas [6] which are the commonest malignancy in children in some series [7, 8].

Rare causes of lymphadenopathy include Kawasaki's disease [9]. In the developed world this is now a more common as a cause of cardiac disease in childhood than rheumatic fever. In some populations the possibility of AIDS should also be considered. Epicondylar lymph nodes may be found in cat scratch fever which is often a diagnosis that is overlooked. It is important to take a history of the patient's pets [10]!

There is a wide diversity of childhood problems that cause lymphadenopathy and in which imaging alone is not diagnostic. The importance of a good history and clinical examination cannot be overemphasized. US takes some time and the discussion that takes place with the patient and parents is often as useful in diagnosis as the examination itself. If a cause for lymphadenopathy is not apparent on imaging, laboratory or clinical grounds then a biopsy or fine needle aspiration is advisable.

5.5.1.2 Ganglia

These common lesions of joints and tendons most often occur at the wrist, where they commonly arise from the scapholunate joint. There is sometimes a history of trauma but most occur spontaneously. They may be found in many locations related to joints and tendons. The fingers and feet are the most common site (Fig. 5.9). The US appearances are of an anechoic mass with acoustic enhancement behind, signs that demonstrate its cystic nature. This can, however, be less obvious as a characteristic in the near field of the US as they may be very close to the skin. This is due to the relatively poor performance of US in the near field. This artefact varies with different machines. Occasionally a ganglion may contain particulate matter [11]. Compression under the US probe will confirm the fluid nature of the lesion. It is often helpful to compare the mass with an area of known fluid at the same depth, for example a vein. The gain level should be adjusted to a point where the known fluid is just echo-free and then the lesion should be re-examined. A solid but hypoechoic mass will then appear brighter than the known area of fluid. Doppler imaging can also exclude the presence of vessels. One pitfall is when a vascular anomaly with low flow rates appears "cystic" in nature and Doppler only shows the vessels when the distal limb is compressed or exercised.

5.5.1.3 Popliteal Cysts

Popliteal or "Baker's" cysts occur behind the knee. They are an anechoic lesion with acoustic enhancement behind arising between the semimembranosus

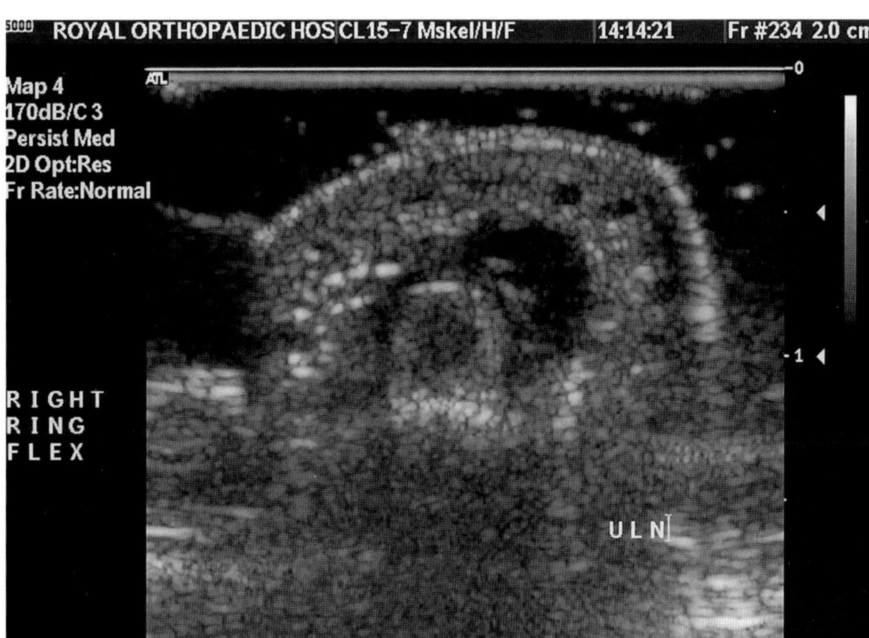

Fig. 5.9 US of a ganglion on the dorsum of the finger

tendon and the medial head of the gastrocnemius muscle. They may be large and can track around the knee even to the anterior regions. The hallmark is a neck or isthmus that runs back to the joint (Fig. 5.10). There is often associated suprapatellar fluid in children [12]. Baker's cysts are less common in children than adults [13] and may contain very thick, jelly-like fluid that is difficult or impossible to aspirate. Chronic lesion are often divided by septa [14, 15].

5.5.1.4 Lipomas

Benign fatty tumours may exhibit many levels of echogenicity (Fig. 5.11) but most commonly they show the same echo pattern as adjacent fat [16]. They should contain fibro-fatty streaks like the adjacent fat. They are usually well defined and displace the surrounding tissue or look like an increased depth of normal fat in comparison to the other side of the

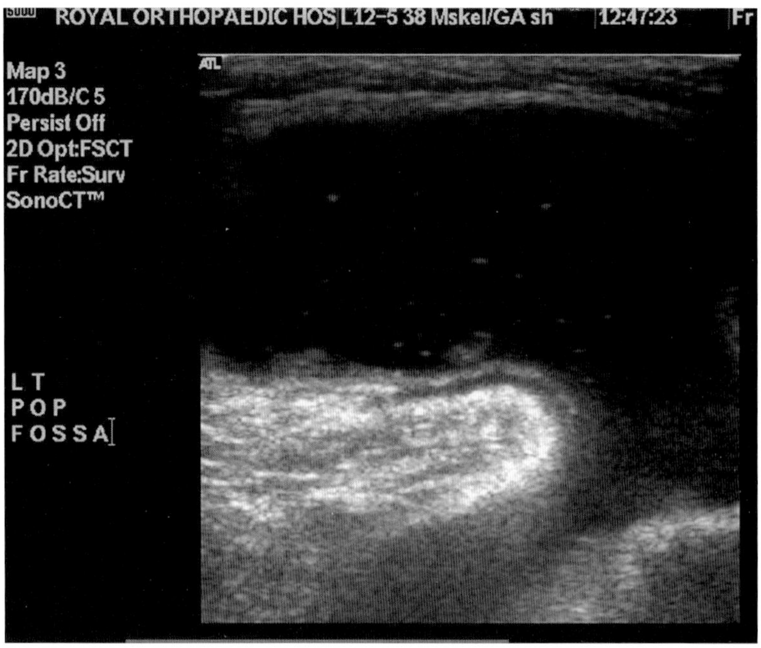

Fig. 5.10 US of a popliteal cyst with the classic "soap bubble" appearance arising from the knee joint

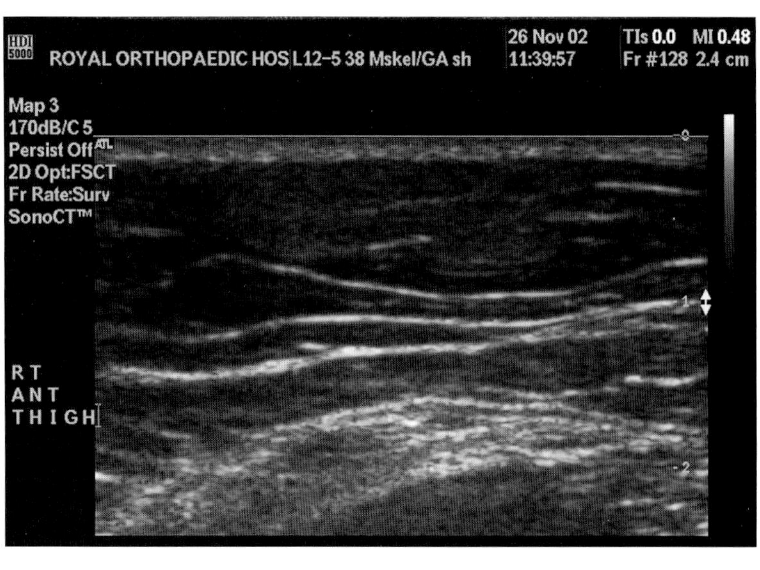

Fig. 5.11 US of a lipoma in the anterior thigh showing uniform hyperechogenicity compared with the surrounding fat

body at the same site. This is an advantage of US imaging as it is simple to compare sides at the same examination with minimal time penalty. There is normally no detectable blood flow in benign lipomas using power Doppler US. Any detectable vascular supply should raise suspicion of malignancy. If there is any doubt, or the history is one of rapid growth, then local staging MRI and a tissue biopsy should be performed [17]. Lipoblastoma is a rare form of "childhood lipoma" that occurs in infancy [18].

5.5.1.5 Sebaceous Cysts

These are cystic structures on US, but may contain some echoes; they are located just underneath the skin. There is usually a detectable punctum clinically and the cysts poke up to the surface (Fig. 5.12). They are avascular which can help the differentiation from skin metastasis which are rare in children [19].

5.5.1.6 Verrucas

Plantar and palmar verrucas are highly vascular lesions of low echogenicity extending with a flat base to the skin surface. They have typical clinical appearances but may be confusing if they are large or in unusual locations [20]. Blood flow is typically increased on Doppler imaging in the immediate surrounding tissues.

5.5.1.7 Foreign Bodies

All types of foreign body will be echogenic but they may be very small. Fortunately, those that are causing symptoms will have produced a local inflammatory reaction which is readily seen on US. The appearances are of an echogenic entity surrounded by an area of low echogenicity. If the foreign material is located in the fingers it may induce a tendinopathy (Fig. 5.13). This occurs within a week or so of the inoculation. The decreased echogenicity around the lesion looks like a "halo" and is due to the foreign body granulation reaction [21]. In a finger it may cause an isolated tenosynovitis rather than a peripheral reaction. Foreign body inoculation is not always remembered by the patient especially in children who may not notice the event as they are so preoccupied with "playing" outdoors.

Wood splinters are a common occurrence in children and will not be demonstrated on plain radiographs. In the initial post-injury phase they may also not be seen with US. If there is a definite injury and removal of the whole of the foreign body is not certain, then it is better to see the child a week after the injury to look for foreign body inoculation with US. By then the classic appearance will be apparent as outlined above. No other method of imaging is as useful in finding retained foreign bodies.

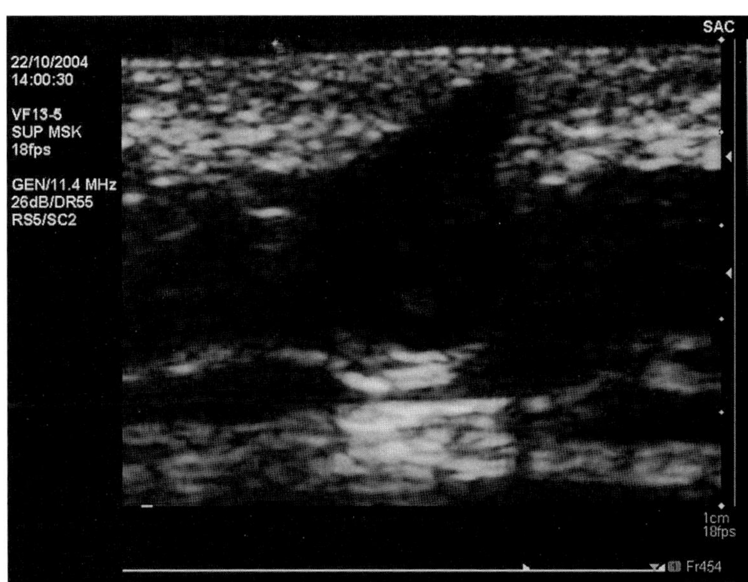

Fig. 5.12 US of a sebaceous cyst which presented as a "lump" showing a punctum extending to the skin

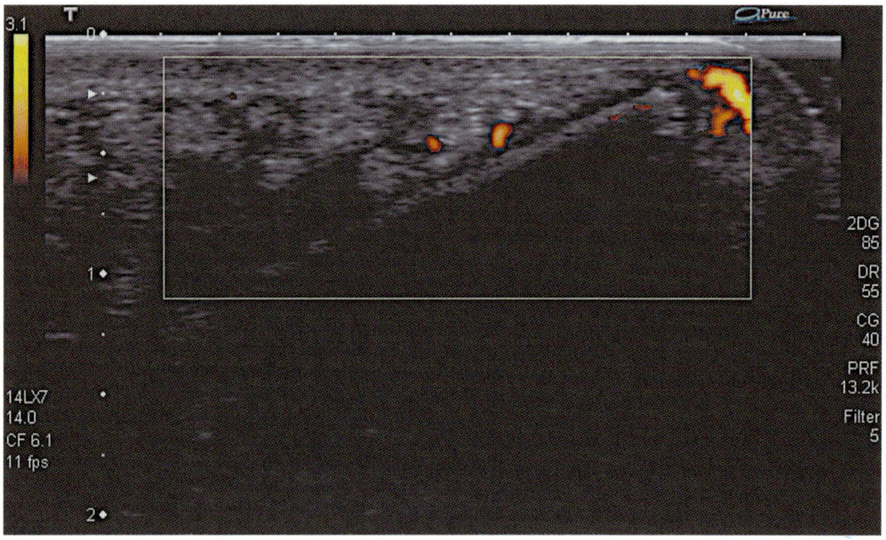

Fig. 5.13 A wooden splinter in the palm of the hand has excited a florid vascular response in the adjacent tissues

5.5.2 Vascular Anomalies

Vascular anomalies are one of the more common lesions in children [22]. The classification of these lesions is complex. They can arise from blood vessels or lymphatic channels. The elements that proliferate may arise from the smooth muscle or endothelium of the vessel. Imaging can determine the flow rate of such vessels and therefore imaging classifications are based on slow or fast "flow".

We divide the abnormalities into (1) haemangiomas and (2) vascular malformations.

1) Haemangiomas

   These are lined with endothelium and appear shortly after birth, growing rapidly in their proliferative phase and involuting over time (Fig. 5.14).

   They are divided histologically into infantile, capillary and cellular types.

   Congenital haemangiomas are present at birth and involute over time [23].

   Other rare vascular tumours include infantile haemangiopericytoma, spindle cell haemangioendothelioma and kaposiform haemangioendothelioma.

   Kasabach-Merritt syndrome is associated with the last two lesions and thrombocytopenia and anaemia with disorders of clotting [24].

2) Vascular anomalies

   The vascular endothelium is stable in these lesions and they are made up of arteries, veins, capillaries, lymphatics and a combination of all of these. They are usually sporadic in appearance but can be associated with genetic disorders. These are not often present at birth but become apparent as the child develops. They are often characterized according to the internal flow rate.

   Fast-flowing lesions are arteriovenous malformations (Fig. 5.15) and fistulas, and slow-flowing lesions are venous, capillary and lymphatic in composition.

   The most well known lymphatic malformation is the cystic hygroma which occurs most commonly in the neck and axilla. These show large fluid-filled spaces that have no flow on US.

   Vascular anomalies are associated with a variety of conditions including: Maffucci's syndrome which has venous malformations, lymphangiomas and multiple exostosis and enchondromas (described by Maffucci in 1881).

   Klippel–Trénaunay syndrome which as well as having a port-wine stain (or capillary malformation) has lymphatic abnormalities with lymphangiomas and lymphatic hypoplasia and varicosities (described by Klippel and Trénaunay in 1900).

   Parkes–Weber syndrome has a capillary naevus with arteriovenous fistulas and varicosities (described by Weber in 1918).

   Proteus syndrome has a capillary naevus with lipohaemangiomas, lipomas, epidermal naevi, lymphangiomas, intraabdominal lipomatosis and partial gigantism with hypertrophy of the hands or feet and asymmetric macrocephaly [25].

   Blue rubber bleb naevus syndrome has involvement of the gastrointestinal tract and skin with venous haemangiomas [26].

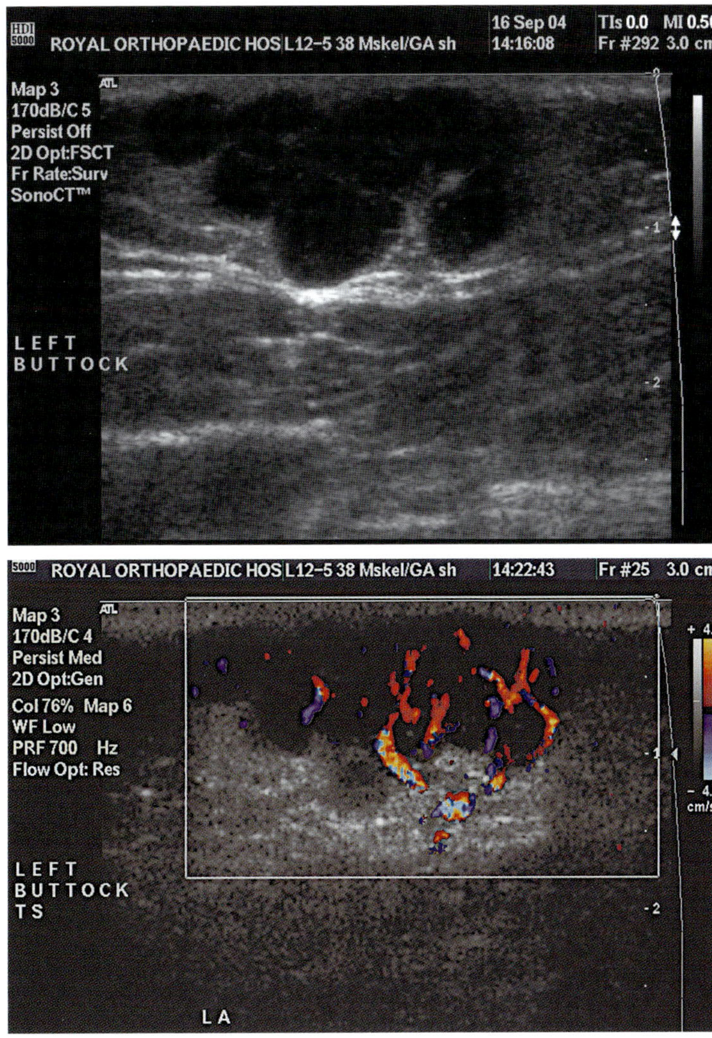

Fig. 5.14a,b A haemangioma in an 18-month-old child which was not evident at birth but is growing "rapidly": (**a**) US appearance, (**b**) US with colour Doppler showing a little flow. The patient also had a visible purple skin blemish

Superficial capillary malformations cannot be seen on MRI and are just noted as an area of increased subcutaneous fat. They are also associated with Sturge-Weber syndrome [27] which has more significant structural abnormalities of the brain.

Doppler signal on US will be dependent on the flow of blood within a lesion. It will show an arterial waveform if of high flow [28]. Sometimes if the blood flow is low, then compression of the probe on the skin or of the distal limb may be needed to confirm the vascularity. Colour Doppler will show the presence of large feeding vessels and at what depth the lesion lies. Superficial vascular lesions will give a bluish hue to the skin. There may be areas of calcification due to phleboliths and these will be detected on US as highly reflective areas with a little acoustic shadowing behind. These are typically seen in haemangiomas.

In small children in whom these lesions are the most common, US is also the easiest imaging to perform, with no need for sedation. If the child cries during the examination this can be an added bonus as the flow through a vascular lesion can be enhanced!

5.5.2.1 Infection

In bone infection a periosteal reaction may be seen in the early phases of osteomyelitis when little is visible by other imaging. However, the opposite is not true; early infection does not always produce a demonstrable periosteal elevation. An abscess can identified as a fluid collection. Although the lesion may contain "solid" echoes, it is well circumscribed and the contents can be seen to swirl especially if the area is compressed. A sinus may be seen as a low echo track between areas of abnormal tissue [29].

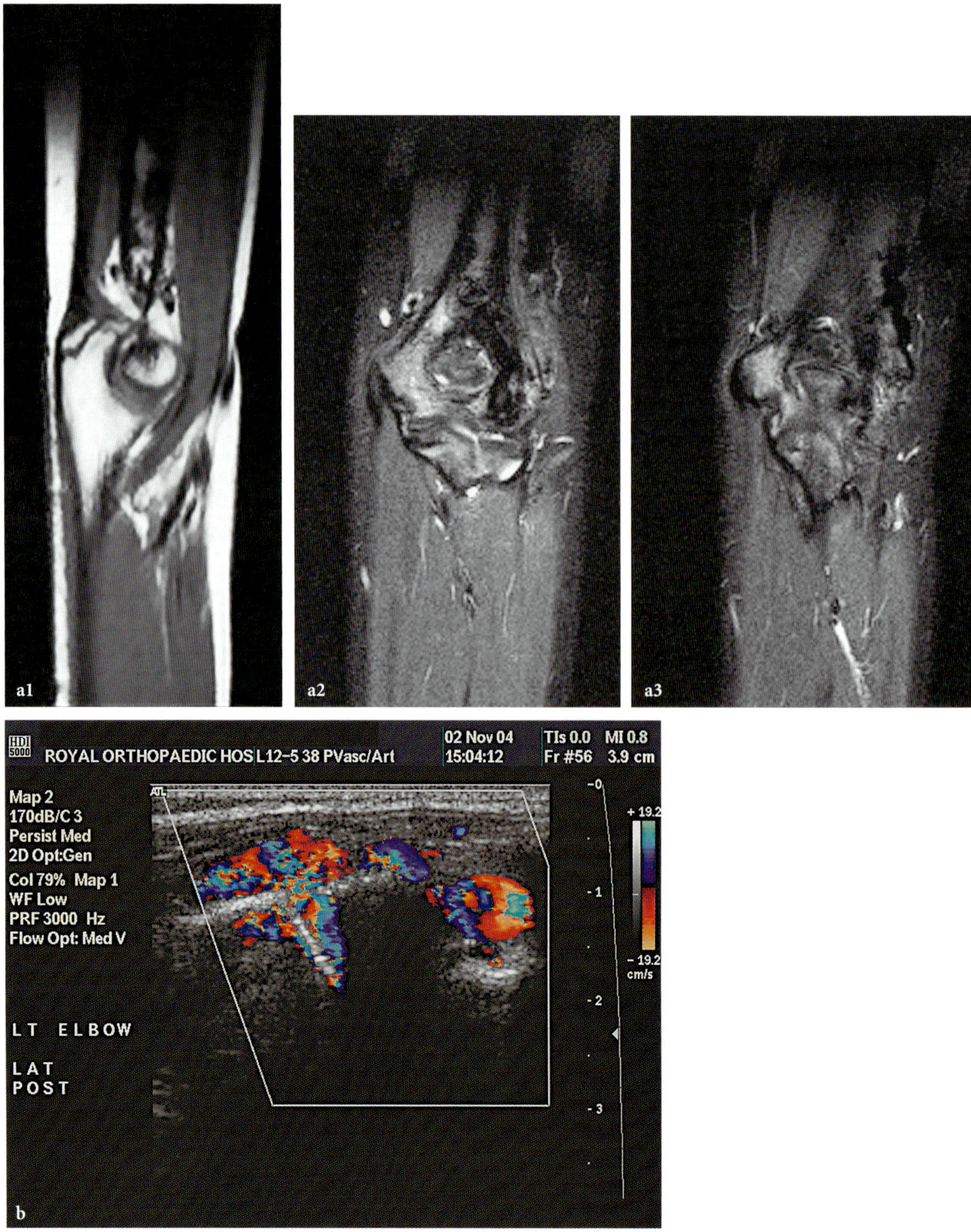

Fig. 5.15a,b An AVM (**a**) seen on MRI to be affecting the bone and (**b**) on US showing high-flow feeding vessels from the soft tissue

5.5.2.2 Muscle Hernias

Muscle hernias present with a lump which is not always palpable. This often occurs when the patient has an MRI examination as the patient is placed in the supine position and the lump disappears. They are much easier to identify with US as the patient can be examined in the standing position and they can show where the lesion is. The author has even had patients whose lumps are only visible on standing after a run just prior to the US study (Fig. 5.16). There is great relief to both the family and patient when a definite diagnosis can be made, and for this problem only US will give the answer [30]!

5.5.2.3 Malignant Lesions

Malignant soft tissue tumours are rare in childhood. There are approximately 100 benign lesions to 1 malignant lesion. The most common soft tissue sarcoma is the rhabdomyosarcoma, and second is the synovial sarcoma (Fig. 5.17) [31].

Rhabdomyosarcomas can arise in any almost organ other than bone. They are derived from primitive mesenchymal tissue which probably has an association with skeletal muscle embryogenesis. Synovial sarcoma, despite its name, is unrelated to the synovium of joints and can be found anywhere in the body, but most commonly in the lower extremities. These tumours are also derived from primi-

a

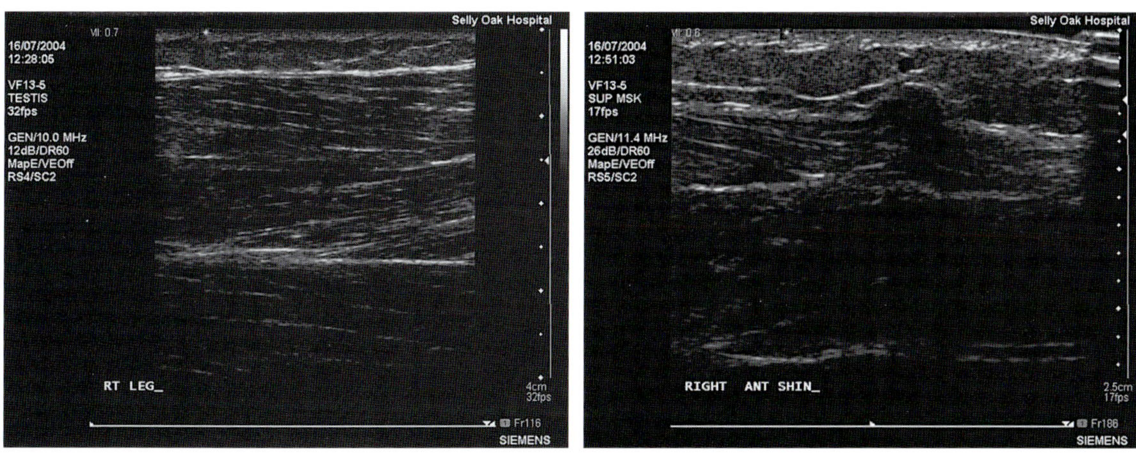

b

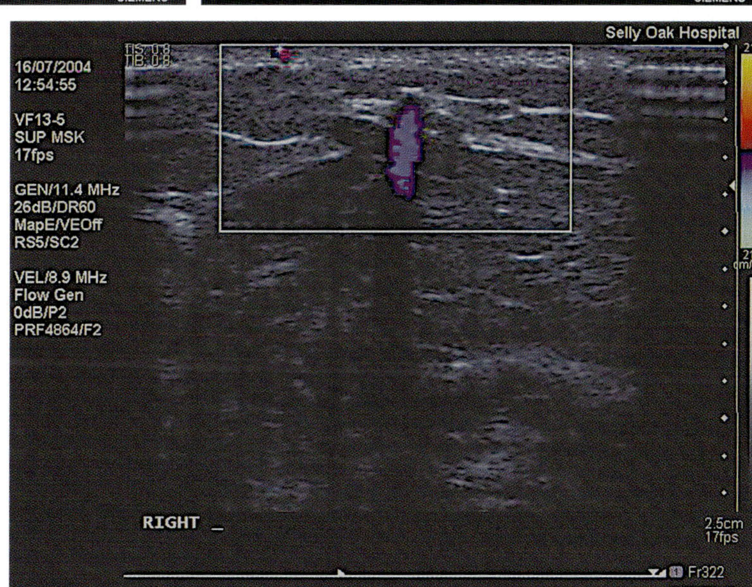

c

Fig. 5.16a,b Muscle hernias: (**a**) normal muscle on scanning the patient in the supine position, (**b**) muscle hernia on US scanning the patient in the standing position, (**c**) colour Doppler US showing the fascial hernia to contain a perforating vein

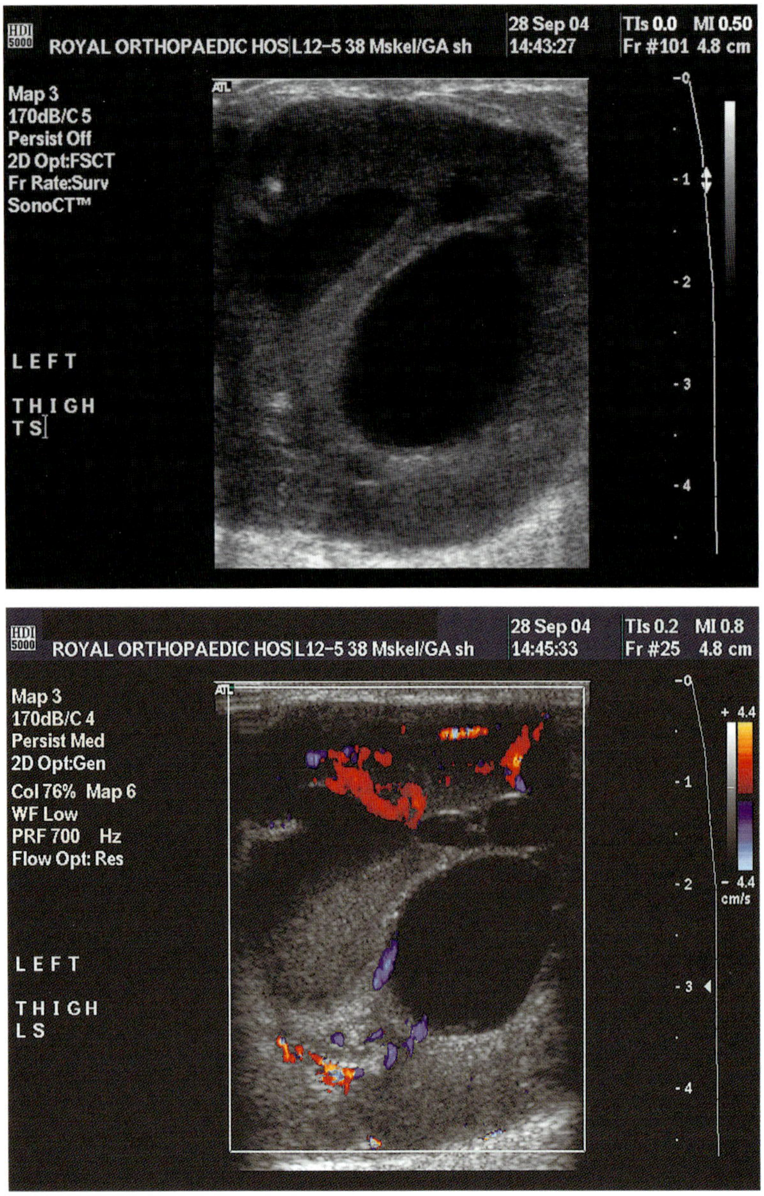

a

b

Fig. 5.17a,b US with colour Doppler of a synovial sarcoma recurrence at the site of previous excision in an 18-year-old, 5 years after the original resection

tive mesenchymal tissue. The bone lesion that can cause soft tissue swelling is the soft tissue extension of a Ewing's sarcoma. Liposarcomas and malignant peripheral nerve sheath tumours are rare.

On US malignant lesions are of variable echogenicity, usually with bizarre vessels, but the evidence of abnormal vascularity alone cannot determine whether a lesion is benign or malignant. They are solid lesions and therefore have a mixed echo pattern. They may contain calcification and then they have "bright" echoes within them. They may also have "cystic" areas which are due to necrosis. US will show the margins and show neurovascular invasion [32] but will not be as useful as MR in providing local staging which is essential for surgical planning. US is used in the assessment of the cartilage cap in osteochondromas especially in the rarer childhood forms of Ollier's disease and Maffucci's syndrome where there are multiple lesions. When the cartilage cap is greater than 3 cm in a child then there is an increased suspicion of malignant transformation into a chondrosarcoma [33].

US can be used to biopsy such a lesion, but once the staging has been completed. This is not only possible in soft tissue lesions but in cases with bone tumours that exhibit extraosseous extension [34, 35].

Liposarcoma is a rare lesion in childhood. They are surprisingly avascular on imaging. The history of rapid growth and a deeper lesion should be more worrying to the clinician [36]. If a "lipoma" is large, increases in size, causes pain, invades muscle or is heterogeneous, then malignancy should be suspected. Any large lesion on US that does not fulfil all the criteria given in the lipoma section above should be imaged with MR and a biopsy guided by US should be undertaken.

Metastasis from endocrine neuroblastoma and renal nephroblastoma (Wilms' tumours) are most common. They are usually in bone but they may have soft tissues extension. Their appearance vary and there may be no discriminating features. The staging should be performed by MR and biopsy. Image guidance may be by fluoroscopy or CT. When there is soft tissue extension or a cortical defect, US guidance is particularly effective.

Nerve tumours include neurofibromas and schwannomas. A neurofibroma is a lesion of low echogenicity. It may have a characteristic "ring" or target sign with an area of higher echogenicity within the lower echogenicity of the outer ring [37] due to the interface of the hypoechoic tumour and the hyperechoic nerve on the inside. The excellent resolution of US can define the nerve from which these lesions arise. If the gain settings are too low a neural tumour may look like a cyst (with acoustic enhancement behind). The method of setting the gain on an area of known fluid as described above should always be used (Fig. 5.18). Children with type 1 and less commonly type 2 neurofibromatosis have multiple neurofibromas.

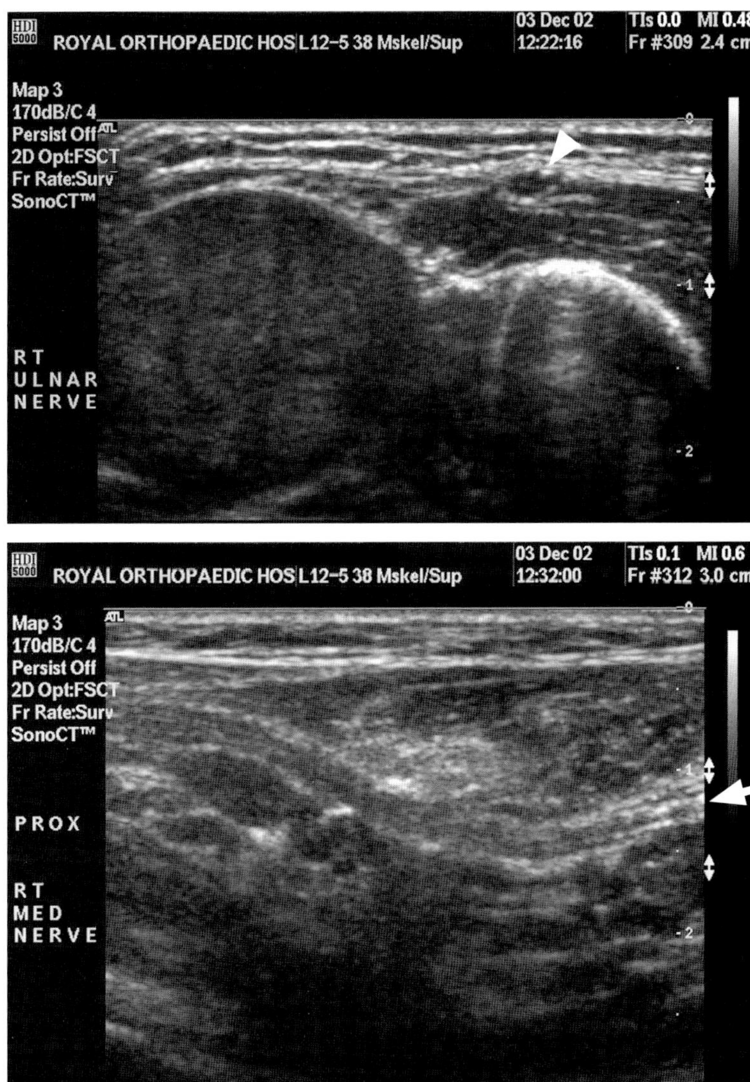

Fig. 5.18a,b A large neurofibroma in a 6-year-old boy with neurofibromatosis type I. **a** Large lesion seen on US with a separate ulnar nerve. **b** US of the median nerve seen longitudinally with the tumour running around it

Schwannomas can be very large and then show areas of "cystic" degeneration which are evident on US. These are less common in children than adults and again are associated with neurofibromatosis [38].

The importance of US in the initial diagnosis of soft tissue malignancy is to determine whether a lesion is solid and then to define those solid lesions where are clear diagnosis is possible using US alone. US is useful to guide biopsy once staging with MR has been performed. This chapter illustrates a variety of lesions that may be assessed and analysed by imaging and where US has an important role. However, the list is not exhaustive and there are rare diagnoses that may benefit from US assessment that are not covered in this text. The same principles apply and the above descriptions should assist the examiner who is confronted by an unusual disease. For details of such disorders the reader is referred to texts on soft tissue tumours [39]. An algorithm for the diagnostic imaging of a soft tissue lump in a child is presented in Fig. 5.19.

5.6 Potential Developments

One group has reported the potential for looking at colour Doppler in tumours to assess response to chemotherapy. They showed a reduction in the colour Doppler signal in those patients who showed a good response to chemotherapy, so perhaps this could be used to assess chemotherapy preoperatively [40].

The follow-up of sarcomas and lymph node involvement has always been difficult. The introduction of positron emission tomography is causing great excitement and may be useful in assessing the extent of malignant lymph node involvement and the response to chemotherapy in such patients [41].

References and Further Reading

1. AbiEzzi SS, Miller LS (1995) The use of ultrasound for the diagnosis of soft-tissue masses in children. J Pediatr Orthop 15(5):566–573
2. Laffan EE, O'Connor R, Ryan SP, et al (2004) Whole-body magnetic resonance imaging: a useful additional sequence in paediatric imaging. Pediatr Radiol 34(6):472–480
3. Stramare R, Tregnaghi A, Fitta C, et al (2004) High-sensitivity power Doppler imaging of normal superficial lymph nodes. J Clin Ultrasound 32(6):273–276
4. Steinkamp HJ, Wissgott C, Rademaker J, et al (2002) Current status of power Doppler and color Doppler sonography in the differential diagnosis of lymph node lesions. Eur Radiol 12(7):1785–1793
5. Rubaltelli L, Proto E, Salmaso R, et al (1990) Sonography of abnormal lymph nodes in vitro: correlation of sonographic and histologic findings. AJR Am J Roentgenol 155(6):1241–1244
6. Moore SW, Schneider JW, Schaaf HS (2004) Diagnostic aspects of cervical lymphadenopathy in children in the developing world: A study of 1877 surgical specimens. Pediatr Surg Int 19:240-244, (June), 2003. J Pediatr Surg 39(7):1150

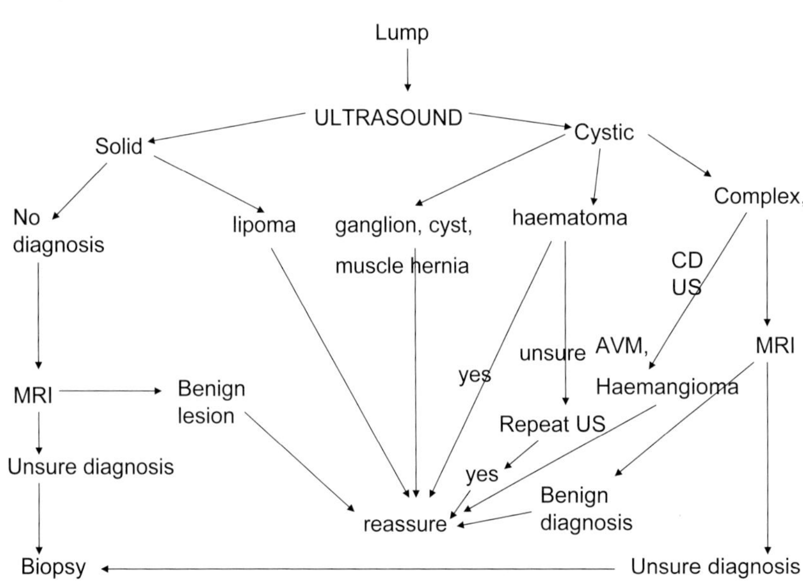

Fig. 5.19. Algorithm for imaging a soft tissue lump in a child

7. Desandes E, Lacour B, Sommelet D, et al (2004) Cancer incidence among adolescents in France. Pediatr Blood Cancer 43(7):742–748
8. Sharma S, Mishra K, Agarwal S, et al (2004) Solid tumors of childhood. Indian J Pediatr 71(6):501–504
9. Newburger JW, Fulton DR (2004) Kawasaki disease. Curr Opin Pediatr 16(5):508–514
10. Lamps LW, Scott MA (2004) Cat-scratch disease: historic, clinical, and pathologic perspectives. Am J Clin Pathol [Suppl] 121:S71–80
11. Ortega R, Fessell DP, Jacobson JA, et al (2002) Sonography of ankle ganglia with pathologic correlation in 10 pediatric and adult patients. AJR Am J Roentgenol 178(6):1445–1449
12. Szer IS, Klein-Gitelman M, DeNardo BA, et al (1992) Ultrasonography in the study of prevalence and clinical evolution of popliteal cysts in children with knee effusions. J Rheumatol 19(3):458–462
13. Seil R, Rupp S, Jochum P, et al (1999) Prevalence of popliteal cysts in children. A sonographic study and review of the literature) Arch Orthop Trauma Surg 119(1–2):73–75
14. Massari L, Faccini R, Lupi L, et al (1990) Diagnosis and treatment of popliteal cysts. Chir Organi Mov 75(3):245–252
15. Lang IM, Hughes DG, Williamson JB, et al (1997) MRI appearance of popliteal cysts in childhood. Pediatr Radiol 27(2):130–132
16. Fornage BD, Tassin GB (1991) Sonographic appearances of superficial soft tissue lipomas. J Clin Ultrasound 19(4):215–220
17. Inampudi P, Jacobson JA, Fessell DP, et al (2004) Soft-tissue lipomas: accuracy of sonography in diagnosis with pathologic correlation. Radiology 233(3):763–767
18. O'Donnell KA, Caty MG, Allen JE, et al (2000) Lipoblastoma: better termed infantile lipoma? Pediatr Surg Int 16(5–6):458–461
19. Giovagnorio F, Valentini C, Paonessa A (2003) High-resolution and color doppler sonography in the evaluation of skin metastases. J Ultrasound Med 22(10):1017–22; quiz 1023–1025
20. Kent H (1969) Warts and ultrasound. Arch Dermatol 100(1):79–81
21. Peterson JJ, Bancroft LW, Kransdorf MJ (2002) Wooden foreign bodies: imaging appearance. AJR Am J Roentgenol 178(3):557–562
22. Laor T (2004) MR imaging of soft tissue tumors and tumor-like lesions. Pediatr Radiol 34(1):24–37
23. Bruckner AL, Frieden IJ (2003) Hemangiomas of infancy. J Am Acad Dermatol 48(4):477–493; quiz 494–496
24. Alvarez-Mendoza A, Lourdes TS, Ridaura-Sanz C, et al (2000) Histopathology of vascular lesions found in Kasabach-Merritt syndrome: review based on 13 cases. Pediatr Dev Pathol 3(6):556–560
25. Dragieva G, Stahel HU, Meyer M, et al (2003) Proteus syndrome. Vasa 32(3):159–163
26. Nahm WK, Moise S, Eichenfield LF, et al (2004) Venous malformations in blue rubber bleb nevus syndrome: variable onset of presentation. J Am Acad Dermatol 50(5 Suppl): S101–106
27. Baselga E (2004) Sturge-Weber syndrome. Semin Cutan Med Surg 23(2):87–98
28. Paltiel HJ, Burrows PE, Kozakewich HP, et al (2000) Soft-tissue vascular anomalies: utility of US for diagnosis. Radiology 214(3):747–754
29. Robben SG (2004) Ultrasonography of musculoskeletal infections in children. Eur Radiol Jan 30 (Epub ahead of print)
30. Beggs I (2003) Sonography of muscle hernias. AJR Am J Roentgenol 180(2):395–399
31. Siegel MJ (2001) Magnetic resonance imaging of musculoskeletal soft tissue masses. Radiol Clin North Am 39(4):701–720
32. Saifuddin A, Burnett SJ, Mitchell R (1998) Pictorial review: ultrasonography of primary bone tumours. Clin Radiol 53(4):239–246
33. Woertler K, Lindner N, Gosheger G, et al (2000) Osteochondroma: MR imaging of tumor-related complications. Eur Radiol 10(5):832–840
34. Rubens DJ, Fultz PJ, Gottlieb RH, et al (1997) Effective ultrasonographically guided intervention for diagnosis of musculoskeletal lesions. J Ultrasound Med 16(12):831–842
35. Torriani M, Etchebehere M, Amstalden E (2002) Sonographically guided core needle biopsy of bone and soft tissue tumors. J Ultrasound Med 21(3):275–281
36. Miller GG, Yanchar NL, Magee JF, et al (1998) Lipoblastoma and liposarcoma in children: an analysis of 9 cases and a review of the literature. Can J Surg 41(6):455–458
37. Beggs I (1998) The ring sign: a new ultrasound sign of peripheral nerve tumours. Clin Radiol 53(11):849–850
38. Beggs I (1997) Pictorial review: imaging of peripheral nerve tumours. Clin Radiol 52(1):8–17
39. De Schepper AM (ed) (2001) Imaging of soft tissue tumours, 2nd edn. Springer, Berlin Heidelberg
40. Bramer JA, Gubler FM, Maas M, et al (2004) Colour Doppler ultrasound predicts chemotherapy response, but not survival in paediatric osteosarcoma. Pediatr Radiol 34(8):614–619
41. Torabi M, Aquino SL, Harisinghani MG (2004) Current concepts in lymph node imaging. J Nucl Med 45(9):1509–1518

6 Interventional Techniques

David Wilson

CONTENTS

6.1 Introduction 85
6.2 Biopsy 85
6.2.1 Soft Tissue Masses 85
6.2.1.1 Consent 85
6.2.1.2 Preparation 86
6.2.1.3 Guidance 86
6.2.1.4 Post-procedure 87
6.2.2 Bone Masses 87
6.2.2.1 Needles 87
6.3 Aspiration 88
6.4 Local Anaesthetic Blocks 89
6.5 Osteoid Osteoma Ablation 89
References and Further Reading 90

6.1 Introduction

Image-guided interventional techniques have the great advantages of limiting the extent of tissue damage, reducing the need for anaesthesia and shortening the stay in hospital. Whilst most of the procedures listed are performed in adults using sedation, it is common practice in children to perform a light general anaesthetic or at least to administer a heavy sedative. Sedation in children can be difficult and hazardous, and we strongly recommend that the procedure is performed under the supervision of a specialized paediatric anaesthetist.

6.2 Biopsy

It is inevitable that soft tissue and bone biopsies will be required in children. The common circumstances are in suspected tumours of bone or soft tissue and when the nature and type of infection is in doubt. In general, there are no major differences from biopsies performed for adults, but there will be many more occasions where a general anaesthetic is necessary.

6.2.1 Soft Tissue Masses

A reasonable approach to soft tissue masses is to determine their nature with ultrasound (US):

- Fluid, solid or mixed
- Vascular or not
- Located in subcutaneous tissues or deeper.

MRI is then important for the lesions that are solid or mixed when the diagnosis is therefore in doubt. From the imaging the biopsy may be planned. There should be formal consultation with the surgeon who would remove the lesion if it proves to be malignant and the pathologist who will interpret the biopsy. Open biopsy will be preferred when there is risk of sampling errors and where the lesion is small and an excision for symptomatic reasons is inevitable.

For many lesions a percutaneous image-guided biopsy will be appropriate. The procedure should include consent, preparation, guidance and post-procedure management.

CT or US may be used to place needles next to a mass that is to be removed surgically. This is especially useful for a small lesion that might be difficult to locate during the operation [1].

6.2.1.1 Consent

Parental consent is mandatory, but it is wise to include the child in the process asking for example "is it alright if I ask your parents permission to do this?"

D. Wilson, FRCP, FRCR
Department of Radiology, Nuffield Orthopaedic Centre, NHS Trust, Windmill Road, Headington, Oxford OX3 7LD, UK

All concerned should be aware that the results of biopsies often take several days to allow time for laboratory analysis and discussion between specialists.

6.2.1.2 Preparation

Although most children will not be at risk from coagulation defects, if there is doubt then coagulation studies should be performed. The room should be quiet and the minimum of staff present. It is wise to allow a parent to accompany the child but the parent should be prepared for the nature of the procedure by discussion separately from their child and they should be seated. It is wise to ask one member of the medical team to be aware that the parent may need support and care.

If general anaesthesia is used then it is still wise to use local anaesthetic to reduce discomfort after the procedure.

6.2.1.3 Guidance

Image guidance will depend on the location of the lesion. It should permit visualization of the area or abnormality and any structure that should be avoided. For example, if there is risk of puncturing bowel, CT is the only safe way of guiding the needle. Most soft tissue masses will be best biopsied using US guidance.

6.2.1.3.1 CT

Has the advantage that the needle is clearly seen and structures to be avoided are apparent [2, 3]. Its disadvantages are that the needle must enter in the plane of scanning and oblique approaches are difficult if not impossible. Also there is a lag between moving the needle and obtaining the image which may be a risk and will prolong the procedure. The radiation dose will mount which may be a particular problem in children.

6.2.1.3.2 US

US allows the direct visualization of the needle as it moves [4–6]. If the needle is at 90° to the US beam it is especially clear. Lesions in limbs are especially easy to biopsy with US guidance as the needle may enter on the side of the limb whilst the probe is held on the top. This means that the probe and jelly do not need to be sterile. When the needle must be placed alongside the probe a sterile cover and sterile jelly are used. The needle tip may be the only part seen as sound reflects off the obliquely placed needle shaft away from the imaging area. Moving the needle slightly will show the tip of the needle as a bright oscillating object. Care should be take to keep the US plane pointing along the needle track or the tip may be lost. If sight of the needle is lost it is best to ignore the screen for a moment and reposition the probe by looking at the

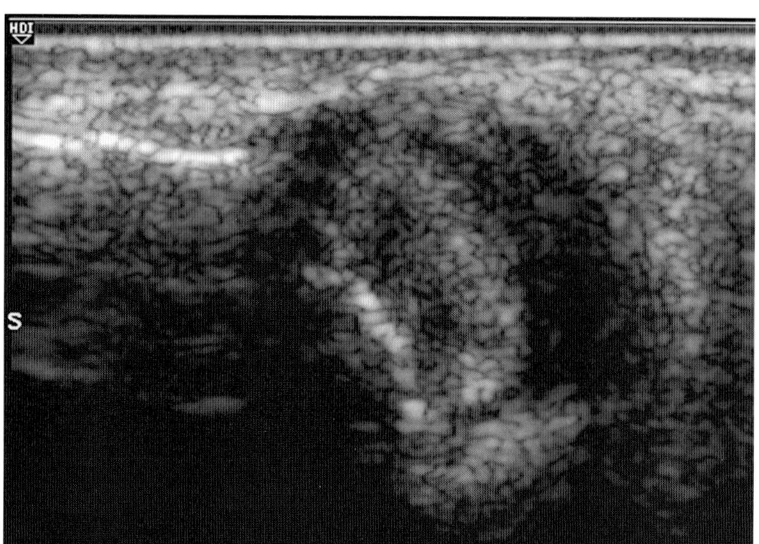

Fig. 6.1 US-guided needle placement next to a tendon thereby avoiding damage to the tendon itself. The needle is introduced at close to 90° to the ultrasound beam allowing visualization of the shaft

patient and the needle. Returning to look at the screen the position will be recaptured (Fig. 6.1).

6.2.1.3.3
MRI

MR has the potential attractions of being free from radiation and allowing the operator to stand next to the patient although an open system is far preferred for this purpose [7–11]. Needles can be seen on MR, although their conspicuousness depends on the alignment with respect to the magnetic field. Interventional MR systems will be available where the track of the needle is predicted by a set of video cameras that locate the needle in space by white makers placed on a needle holding extension. Rapid re-imaging with say 1 second refreshing will then allow the needle to be followed. The needle and all equipment will need to be MRI-safe. These needles tend to be expensive.

With all imaging a side-cutting needle is most effective for soft tissue biopsies (Fig. 6.2). It is wise to practice with the needle beforehand. This also helps to warn the patient about the click that spring-loaded systems make. The open side of the needle should be placed in the area of interest and the sheath withdrawn from the area. This means holding the central part still and pulling the outer part backwards. Reversing this action would push the needle beyond the area and should be avoided. At least two specimens should be taken and preferably several. Specimens should be sent for histological diagnosis and for microbiological culture in all cases. (Look at the cell for infection and culture the tumour.) This practice will reduce the risk of repeat biopsy; however sure you are on imaging, mistakes of classification are common. Check beforehand what type of specimen bottle is needed and whether to use fixative; some laboratories prefer unfixed specimens.

6.2.1.4
Post-procedure

Risks of biopsy include, puncture of vessels and viscus, infection, allergy to the drugs and haemorrhage. The time of post-procedure observation will depend on how likely these risks are and the nature of sedation or anaesthesia. Clear written instructions should be given to the ward or day-case unit staff and analgesia should be prescribed.

6.2.2
Bone Masses

The principles outlined above for soft tissue masses all apply to bone lesions. The differences are small but centre around the nature of guidance. US is less appropriate and most will use either fluoroscopy or CT. However, some authors have suggested that cortical defect seen on US will allow effective guidance with this technique [12, 13]. Again the technique depends on seeing the lesion and important intervening structures.

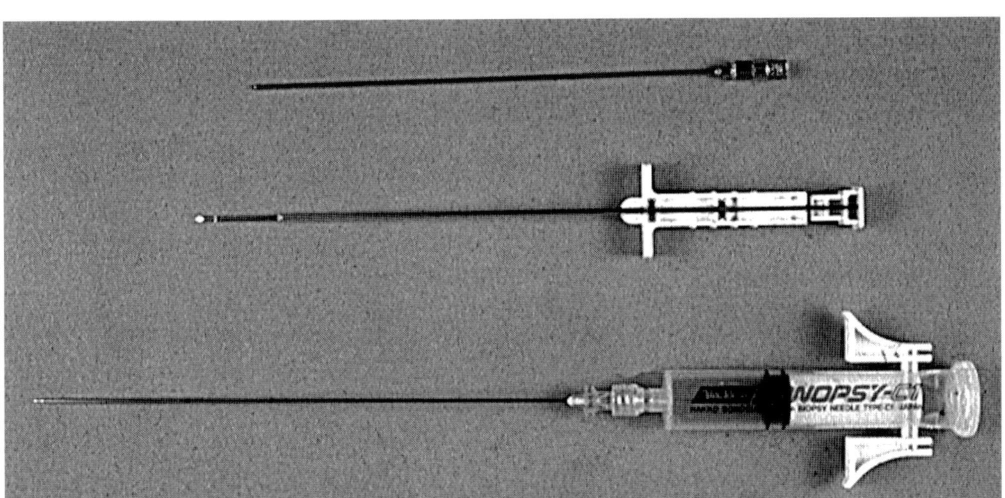

Fig. 6.2 A variety of soft tissue biopsy needles. The side-cutting type is the easiest to use and the most effective

6.2.2.1
Needles

There are several commercially available bone biopsy needles. The two common types are the tapered needle with a trocar and the cannula with a central cutting needle.

The tapered needle traps the bone specimen which must be expelled by pushing from the tip to the hub. This means the needle must be removed and a second specimen requires reinsertion and guidance. There is also the risk of puncturing the operator's hands with the needle tip when expelling the specimen. Non tapered needles have the risk that the specimen may escape. This risk is reduced by wobbling the needle before extraction and by applying gentle suction with a syringe. Strong suction may pull the specimen into the syringe damaging it en route.

The cannula type of needle allows the cutting needle to be inserted through a cannula that has been placed up to the bone surface or the edge of the lesion. Repeat biopsy specimens are then safe and easy. A modification of the cannula allows a drill to be introduced to penetrate hard bone cortex. The drill point is eccentric and this causes the hole to be larger than the drill; the cannula then may be advanced into the drilled hole (Fig. 6.3).

Both types of needle can have a smooth cutting edge or a saw-toothed one. The latter is tougher and enters hard lesions better but may fragment the specimen.

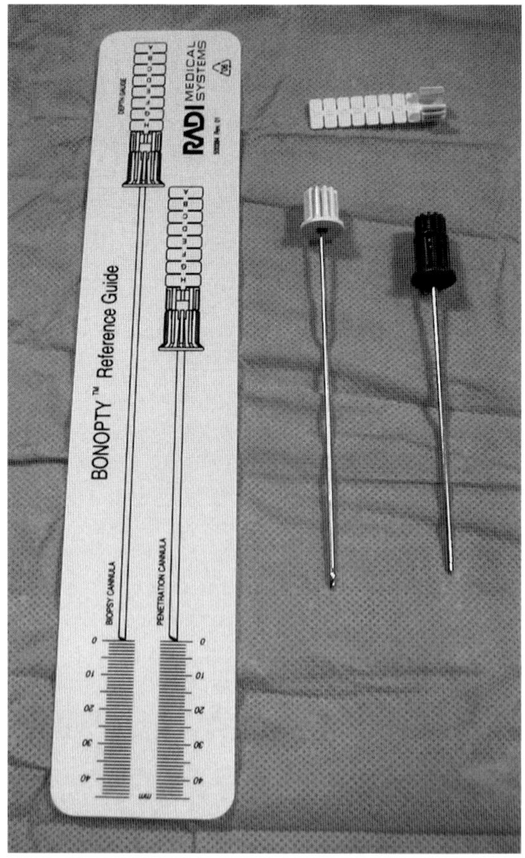

Fig. 6.3 A Boneopty bone biopsy system with an eccentric drill to make a hole larger than the cannula. The outer cannula enters the bone and allows repeated biopsies

6.3
Aspiration

Some joints may be aspirated by puncture guided by palpation and surface landmarks. This is especially true for the knee. However, using US improves the success rate for even the more superficial joints [14, 15]. Deep and complex joints may be difficult to reach and when effusions are small or complicated by extensive synovial thickening there are great advantages in image guidance.

Typical reasons for aspiration are:
- Suspected septic arthritis
- Painful haemarthrosis
- Synovitis
- Symptomatic effusion
- Therapeutic tests for the origin of pain

Therapeutic/diagnostic aspiration of collections of fluid adjacent to bone has been advocated in children in whom sickle cells infarction cannot be differentiated from osteomyelitis. All but one of the collections resulting from infection were greater than 10 mm in depth [16]. It has also been advocated for other more common types of osteomyelitis aspiration/biopsy [17].

Guidance methods include US, fluoroscopy, CT and MRI. The first two are so effective that the more complex methods are virtually never required. The guidance principles are identical to those for soft tissue mass biopsy plus the following suggestions.

Aspiration of the hip is easiest when the child is supine [18]. The site of the greatest capsular distension is marked on the skin vertically above the collection. The US may then be put away as, a direct vertical puncture with a standard venepuncture needle pushed down to the bone is a very reliable method [19].

The majority of joints are of the ball and socket configuration. One side is convex and the other concave. This means that the needle needs to be directed from the convex side into to the concavity. Fluoroscopic projection of the joint space may be misleading as there is often a lip of bone from the concave side overlapping the joint. However, aiming the needle to hit the bone that is convex and then walking it towards the joint gives the desired obliquity to enter. US guidance allows the joint to be seen including any lip and makes this process easier.

To be certain that the joint has been entered when there is no effusion it helps to introduce some non-ionic radiographic contrast agent using fluoroscopy. US is more difficult if local anaesthetic in a syringe connected to the needle flows into the joint; there will then be no local collection seen on US. The injectate will flow easily. For retrospective confirmation of intra-articular injection it is possible to add some radiographic contrast and then take a plain radiograph to follow.

6.4 Local Anaesthetic Blocks

Guidance for therapeutic or diagnostic blocks may be by US, fluoroscopy, CT or MRI depending on location, intervening structures and the operator's expertise. For example, fluoroscopy is most often used for spinal root blocks and US is ideal for painful soft tissue lesions.

6.5 Osteoid Osteoma Ablation

Osteoid osteoma is a benign but very painful tumour of bone that often affects children. It is fairly uncommon but treatment is very effective. Typically the pain is at night and responds dramatically to prostaglandin-blocking drugs such as aspirin. Treatment used to be by surgical excision of the tiny nidus which is a few millimetres in diameter. It is not necessary to excise the sclerotic reaction around the nidus. Recently it has been realized that radiological techniques are just as effective, and surgery is now rarely indicated [20–25]. Methods include the excision of the nidus by a fairly wide bone biopsy needle and thermal ablation by a radiofrequency-heated needle tip and laser ablation (Figs. 6.4, 6.5) [26].

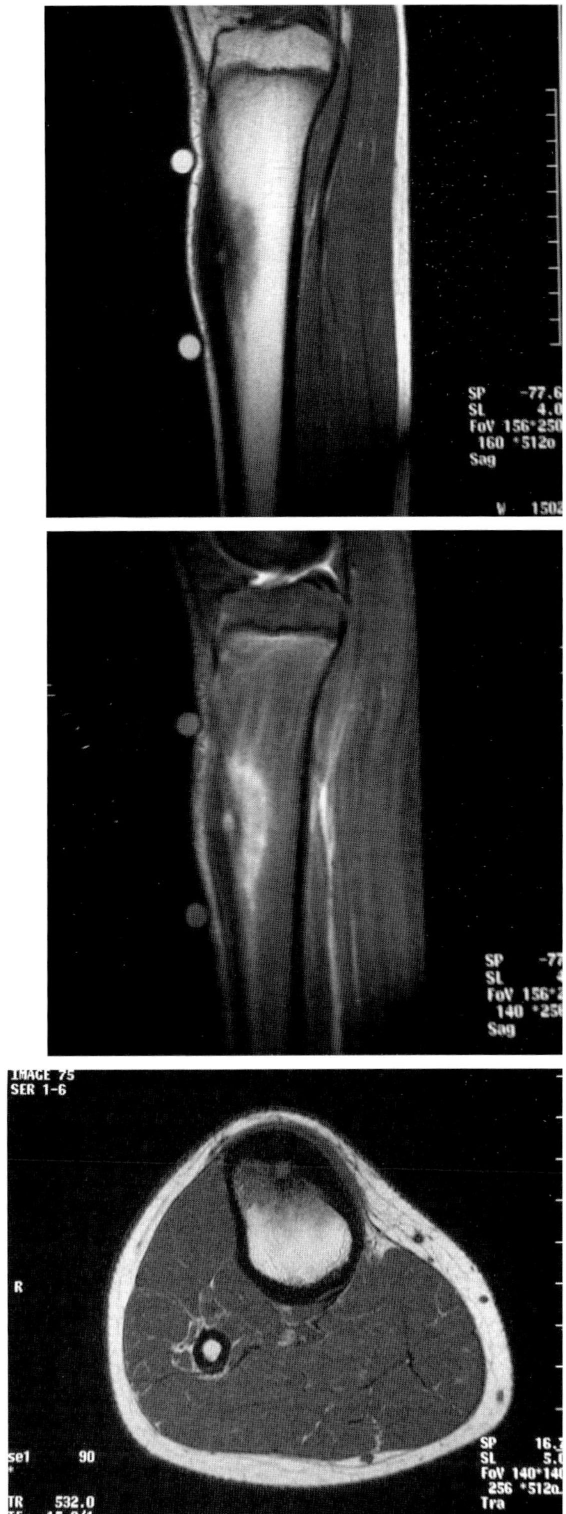

Fig. 6.4 a T1-weighted spin-echo image of the tibia showing an area of oedema and a nidus below the thickened cortex of the tibia. **b** FSTIR image shows a halo of oedema around the osteoid osteoma. **c** Axial T1-weighted spin-echo image confirms the location of the nidus

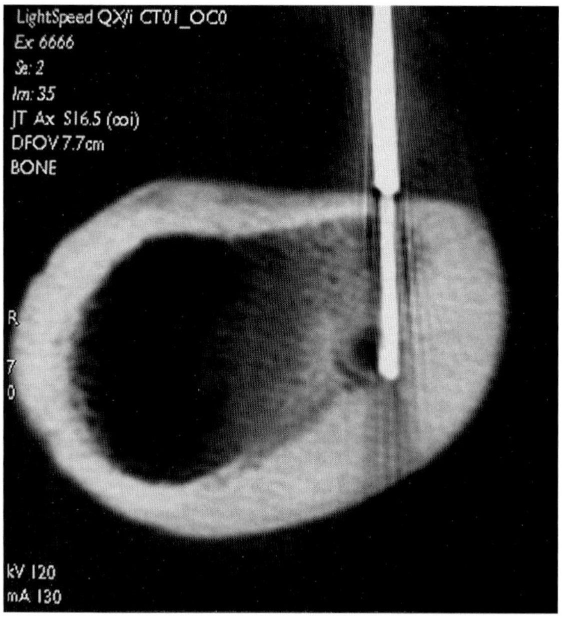

Fig. 6.5 CT-guided placement of the bone biopsy needle prior to radiofrequency ablation

Heating techniques may be a risk if the lesion is near to a nerve, which is often the case when the lamina of a vertebral body is affected. This may be overcome by using saline irrigation of the epidural space or by relying on the simple biopsy method. Image guidance is invariably by CT as the lesions are small and difficult or even impossible to see with other methods.

References and Further Reading

1. Hardaway BW, Hoffer FA, Rao BN (2000) Needle localization of small pediatric tumors for surgical biopsy. Pediatr Radiol 30(5):318–322
2. Hussain HK, Kingston JE, Domizio P, et al (2001) Imaging-guided core biopsy for the diagnosis of malignant tumors in pediatric patients. AJR Am J Roentgenol 176(1):43–47
3. Agid R, Sklair-Levy M, Bloom AI, et al (2003) CT-guided biopsy with cutting-edge needle for the diagnosis of malignant lymphoma: experience of 267 biopsies. Clin Radiol 58(2):143–147
4. Konermann W, Wuisman P, Hillmann A, et al (1995) Value of sonographically guided biopsy in the histological diagnosis of benign and malignant soft-tissue and bone tumors (in German). Z Orthop Ihre Grenzgeb 133(5):411–421
5. Konermann W, Wuisman P, Ellermann A, et al (2000) Ultrasonographically guided needle biopsy of benign and malignant soft tissue and bone tumors. J Ultrasound Med 19(7):465–471
6. Mayekawa DS, Ralls PW, Kerr RM, et al (1989) Sonographically guided arthrocentesis of the hip. J Ultrasound Med 8(12):665–667
7. Schulz T, Bennek J, Schneider JP, et al (2003) MRI-guided pediatric interventions (in German). Rofo 175(12):1673–1681
8. Daecke W, Libicher M, Madler U, et al (2003) MRI-guided musculoskeletal biopsy (in German). Orthopade 32(2):170–174
9. Genant JW, Vandevenne JE, Bergman AG, et al (2002) Interventional musculoskeletal procedures performed by using MR imaging guidance with a vertically open MR unit: assessment of techniques and applicability. Radiology 223(1):127–136
10. Koskinen SK, Parkkola RK, Karhu J, et al (1997) Orthopedic and interventional applications at low field MRI with horizontally open configuration. A review. Radiologe 37(10):819–824
11. Martorano D, Verna V, Mancini A, et al (2003) CT evaluation pre- and post-percutaneous ablation by radiofrequency of osteoid osteoma. Preliminary experience. Chir Organi Mov 88(2):233–240
12. Gil-Sanchez S, Marco-Domenech SF, Irurzun-Lopez J, et al (2001) Ultrasound-guided skeletal biopsies. Skeletal Radiol 30(11):615–619
13. Gupta S, Takhtani D, Gulati M, et al (1999) Sonographically guided fine-needle aspiration biopsy of lytic lesions of the spine: technique and indications. J Clin Ultrasound 27(3):123–129
14. Balint PV, Kane D, Hunter J, et al (2002) Ultrasound guided versus conventional joint and soft tissue fluid aspiration in rheumatology practice: a pilot study. J Rheumatol 29(10):2209–2213
15. Raza K, Lee CY, Pilling D, et al (2003) Ultrasound guidance allows accurate needle placement and aspiration from small joints in patients with early inflammatory arthritis. Rheumatology (Oxford) 42(8):976–979
16. Booz MM, Hariharan V, Aradi AJ, et al (1999) The value of ultrasound and aspiration in differentiating vaso-occlusive crisis and osteomyelitis in sickle cell disease patients. Clin Radiol 54(10):636–639
17. Sammak B, Abd El Bagi M, Al Shahed M, et al (1999) Osteomyelitis: a review of currently used imaging techniques. Eur Radiol 9(5):894–900
18. Alexander JE, Seibert JJ, Glasier CM, et al (1989) High-resolution hip ultrasound in the limping child. J Clin Ultrasound 17(1):19–24
19. Berman L, Fink AM, Wilson D, et al (1995) Technical note: identifying and aspirating hip effusions. Br J Radiol 68(807):306–310
20. Cantwell CP, Obyrne J, Eustace S (2004) Current trends in treatment of osteoid osteoma with an emphasis on radiofrequency ablation. Eur Radiol 14(4):607–617

21. Barei DP, Moreau G, Scarborough MT, et al (2000) Percutaneous radiofrequency ablation of osteoid osteoma. Clin Orthop (373):115–124
22. Rosenthal DI, Hornicek FJ, Torriani M, et al (2003) Osteoid osteoma: percutaneous treatment with radiofrequency energy. Radiology 229(1):171–175
23. Cioni R, Armillotta N, Bargellini I, et al (2004) CT-guided radiofrequency ablation of osteoid osteoma: long-term results. Eur Radiol 14(7):1203–1208
24. Pinto CH, Taminiau AH, Vanderschueren GM, et al (2002) Technical considerations in CT-guided radiofrequency thermal ablation of osteoid osteoma: tricks of the trade. AJR Am J Roentgenol 179(6):1633–1642
25. Venbrux AC, Montague BJ, Murphy KP, et al (2003) Image-guided percutaneous radiofrequency ablation for osteoid osteomas. J Vasc Interv Radiol 14(3):375–380
26. DeFriend DE, Smith SP, Hughes PM (2003) Percutaneous laser photocoagulation of osteoid osteomas under CT guidance. Clin Radiol 58(3):222–226

Subject Index

A
abscess 60
accessory ossicles 34
acoustic enhancement 73
AIDS 73
anaesthesia
- general 86
- local 86
anisotropy 40
apophyseal
- avulsion 19, 21, 22
- injury 21
arteriovenous malformation 7
arthritis
- juvenile idiopathic 43, 49
- juvenile rheumatoid 49
- septic 54
aspiration 57, 88
avulsion injury 43

B
Baker's cyst 73
Barlow
- manoeuvre 2
- stress test 4
biomechanics 19
biopsy 85
blue rubber bleb naevus syndrome 76
bone
- biopsy 87
- scintigraphy 30
- - in irritable hip 54
bowing fracture 20
bursal inflammation 34
butterfly vertebra 11–13

C
calcification 68, 69
cellulitis 60
cervical lymphadenopathy 72
Chiari malformation 12
chondral fracture 23
chondrosarcoma 80
chronic overuse syndrome 20
Cobb angle 12
congenital birth defect 1
consent 85
cord tumour 11
CT
- biopsy 86
- in irritable hip 56
- myelography 14
cyclical injury 20
cystic hygroma 76

D
developmental dysplasia of the hip (DDH) 1
- risk factors 3
diaphyseal injury 20
diastematomyelia 11–13
Doppler ultrasound 6
- osteonecrosis 6
dynamic examination 4
dysplasia 1

E
echondroma 76
effusion 54
eosinophilic granuloma 30
Ewing's sarcoma 30, 80

F
fistula 76
focal defect 7
foramen magnum defect 12
foreign bodies 25, 75
Freiberg's disease 32
fused (block) vertebra 11

G
ganglia 73
golfer's elbow 34
greenstick fracture 20, 24
growth arrest 28

H
haemangioma 68, 76
haematoma 25, 67
hemivertebra 11, 12
Holt–Oram syndrome 8

I
idiopathic kyphosis 11
impingement syndrome 26

infectious mononucleosis 72
injury
- apophyseal 21
- avulsion 43
- cyclical 20
- diaphyseal 20
- metaphyseal 20
- muscle 26
- osteochondral 20, 23
- overuse 19, 30, 43
- physeal 21
- Salter-Harris type 19, 28, 29
- tendon 26
irritable hip 53

J
jumper's knee, see Sinding-Larsen(-Johansson) disease
juvenile
- idiopathic arthritis 43, 49
- rheumatoid arthritis 49

K
Kawasaki's disease 73
kinetic chain 19
Klippel-Trénaunay syndrome 76
Kohler's disease 32
kyphoscoliosis 11

L
ligament 40
- anterior talofibular 50
- anterior tibiofibular 50
- MR of ligament injuries 50
lipoblastoma 75
lipohaemarthrosis 21
lipoma 74
- of the cord 11, 12
liposarcoma 80, 91
little league elbow 34
local anaesthetic block 89
lordoscoliosis 11
lymph node 71
lymphangioma 76

M
Maffucci's syndrome 76, 80
malignant peripheral nerve sheath tumour 80
medial epicondylitis 34
meningocele 12
mesotendon 39
metaphyseal injury 20
metastasis 70
MR
- biopsy 87
- in irritable hip 56
- infection 60
- of ligament injuries 50
- tendon 42
multiple exostosis 76
muscle
- hernia 79
- injury 26
musculotendinous junction 20
myelography 12
myelomeningocele 10, 11, 15
myositis ossificans 68
myotendinous disruption 27

N
needle 88
nephroblastoma (Wilm's tumour) 81
neural
- arch defect 12
- tube defect 10, 11, 14
neurofibroma 81
neurofibromatosis 81

O
Ollier's disease 80
Ortolani
- manoeuvre 2
- stress test 4
Osgood-Schlatter disease 32, 43
osteochondral
- fracture 23
- fragment 21, 23, 24
- injury 20, 23
- lesion 33
osteochondritis dissecans 32
osteochondrosis 28, 32
osteoid osteoma 31, 70
- ablation 89
osteomyelitis 57, 59
- chronic recurrent multifocal 60
osteonecrosis 6
- Doppler ultrasound 6
overuse injury 19, 30, 43

P
Panner's disease 33
pannus 48
Parkes-Weber syndrome 76
patellar sleeve fracture 21-23
periosteal reaction 77
Perthes' disease 54, 57
pes anserinus irritation 34
phocomelia 7
physeal injury 21
proteus syndrome 76

R
radiofrequency ablation 89
rhabdomyosarcoma 79
rotator cuff tendinopathy 26

S
Salter-Harris type injury 19, 28, 29
Scheuermann's disease

schwannoma 81, 82
scintigraphy for infection 60
scoliosis 11, 14
- idiopathic 11
screening 70
sebaceous cyst 75
septic arthritis 54
Sever's disease 32
Sinding-Larsen-Johansson disease (jumper's knee) 32, 33, 43
skip metastasis 69
slipped capital femoral epiphysis 28
snapping hip 46
soft tissue
- infection 59
- lump 82
- - algorithm 82
- masses 85
spina bifida 10
spinal cord tethering 11, 12
splint therapy 4
split cord 12
spondylolysis 31
stress fracture 28, 31
Sturge-Weber syndrome 78
SUFE 54, 57
supracondylar fracture 24
synovial biopsy 48
syringomyelia 11, 12

T
talipes equinovarus 9
tendino-osseous junction 43
tendon
- injury 26
- MR imaging 42
- type 1 39
- type 2 40
tenosynovitis 48, 49
- infectious type 43
tethered cord 15
thermal ablation 89
torus fracture 20, 21, 24
transient synovitis 53

U
ultrasound (US)
- biopsy 86
- Graf technique 3, 4
- greenstick fracture 24
- in irritable hip 54
- infection 60
- mass lesions 67
- Morin technique 4
- screening 3, 5, 6
- torus fracture 24

V
vascular anomaly 76
verruca 75
vertebra
- butterfly 11-13
- fused (block) 11
- hemivertebra 11, 12

W
wedged vertebra 11
Wilm's tumour 81

List of Contributors

Gina Allen, DCH, MRCP, FRCR
The Royal Orthopaedic Hospital NHS Trust
Bristol Road South
Northfield
Birmingham, B31 2AP
UK

Stefano Bianchi, MD
Institut de Radiologie
Cliniques des Grangettes
Fondation des Grangettes
1224 Chene-Bougeries
Switzerland

Ruth Cheung, FRCR
Department of Radiology
Nuffield Orthopaedic Centre, NHS Trust
Windmill Road, Headington
Oxford OX3 7LD
UK

Clare Groves, FRCR
Department of Radiology
The General Infirmary at Leeds
Leeds LS1 3EX
UK

Carlo Martinoli, MD
Cattedra di Radiologia "R"
Università di Genova
Largo Rosanna Benzi 8
16132 Genoa
Italy

Philip J. O'Connor, FRCR
Department of Radiology
The General Infirmary at Leeds
Leeds LS1 3EX
UK

Paolo Tomà, MD
Reparto di Radiologia
Istituto "Giannina Gaslini"
Largo Gaslini 5
16148 Genoa
Italy

Maura Valle, MD
Reparto di Radiologia
Istituto "Giannina Gaslini"
Largo Gaslini 5
16148 Genoa
Italy

David Wilson, FRCP, FRCR
Department of Radiology
Nuffield Orthopaedic Centre
NHS Trust
Windmill Road, Headington
Oxford OX3 7LD
UK

Medical Radiology Diagnostic Imaging and Radiation Oncology

Titles in the series already published

Diagnostic Imaging

Innovations in Diagnostic Imaging
Edited by J. H. Anderson

Radiology of the Upper Urinary Tract
Edited by E. K. Lang

The Thymus - Diagnostic Imaging, Functions, and Pathologic Anatomy
Edited by E. Walter, E. Willich, and W. R. Webb

Interventional Neuroradiology
Edited by A. Valavanis

Radiology of the Pancreas
Edited by A. L. Baert, co-edited by G. Delorme

Radiology of the Lower Urinary Tract
Edited by E. K. Lang

Magnetic Resonance Angiography
Edited by I. P. Arlart, G. M. Bongartz, and G. Marchal

Contrast-Enhanced MRI of the Breast
S. Heywang-Köbrunner and R. Beck

Spiral CT of the Chest
Edited by M. Rémy-Jardin and J. Rémy

Radiological Diagnosis of Breast Diseases
Edited by M. Friedrich and E.A. Sickles

Radiology of the Trauma
Edited by M. Heller and A. Fink

Biliary Tract Radiology
Edited by P. Rossi, co-edited by M. Brezi

Radiological Imaging of Sports Injuries
Edited by C. Masciocchi

Modern Imaging of the Alimentary Tube
Edited by A. R. Margulis

Diagnosis and Therapy of Spinal Tumors
Edited by P. R. Algra, J. Valk, and J. J. Heimans

Interventional Magnetic Resonance Imaging
Edited by J.F. Debatin and G. Adam

Abdominal and Pelvic MRI
Edited by A. Heuck and M. Reiser

Orthopedic Imaging
Techniques and Applications
Edited by A. M. Davies and H. Pettersson

Radiology of the Female Pelvic Organs
Edited by E. K.Lang

Magnetic Resonance of the Heart and Great Vessels
Clinical Applications
Edited by J. Bogaert, A.J. Duerinckx, and F. E. Rademakers

Modern Head and Neck Imaging
Edited by S. K. Mukherji and J. A. Castelijns

Radiological Imaging of Endocrine Diseases
Edited by J. N. Bruneton in collaboration with B. Padovani and M.-Y. Mourou

Trends in Contrast Media
Edited by H. S. Thomsen, R. N. Muller, and R. F. Mattrey

Functional MRI
Edited by C. T. W. Moonen and P. A. Bandettini

Radiology of the Pancreas
2nd Revised Edition
Edited by A. L. Baert
Co-edited by G. Delorme and L. Van Hoe

Emergency Pediatric Radiology
Edited by H. Carty

Spiral CT of the Abdomen
Edited by F. Terrier, M. Grossholz, and C. D. Becker

Liver Malignancies
Diagnostic and Interventional Radiology
Edited by C. Bartolozzi and R. Lencioni

Medical Imaging of the Spleen
Edited by A. M. De Schepper and F. Vanhoenacker

Radiology of Peripheral Vascular Diseases
Edited by E. Zeitler

Diagnostic Nuclear Medicine
Edited by C. Schiepers

Radiology of Blunt Trauma of the Chest
P. Schnyder and M. Wintermark

Portal Hypertension
Diagnostic Imaging-Guided Therapy
Edited by P. Rossi
Co-edited by P. Ricci and L. Broglia

Recent Advances in Diagnostic Neuroradiology
Edited by Ph. Demaerel

Virtual Endoscopy and Related 3D Techniques
Edited by P. Rogalla, J. Terwisscha Van Scheltinga, and B. Hamm

Multislice CT
Edited by M. F. Reiser, M. Takahashi, M. Modic, and R. Bruening

Pediatric Uroradiology
Edited by R. Fotter

Transfontanellar Doppler Imaging in Neonates
A. Couture and C. Veyrac

Radiology of AIDS
A Practical Approach
Edited by J.W.A.J. Reeders and P.C. Goodman

CT of the Peritoneum
Armando Rossi and Giorgio Rossi

Magnetic Resonance Angiography
2nd Revised Edition
Edited by I. P. Arlart, G. M. Bongratz, and G. Marchal

Medical Radiology Diagnostic Imaging and Radiation Oncology

Titles in the series already published

Pediatric Chest Imaging
Edited by Javier Lucaya and Janet L. Strife

Applications of Sonography in Head and Neck Pathology
Edited by J. N. Bruneton in collaboration with C. Raffaelli and O. Dassonville

Imaging of the Larynx
Edited by R. Hermans

3D Image Processing
Techniques and Clinical Applications
Edited by D. Caramella and C. Bartolozzi

Imaging of Orbital and Visual Pathway Pathology
Edited by W. S. Müller-Forell

Pediatric ENT Radiology
Edited by S. J. King and A. E. Boothroyd

Radiological Imaging of the Small Intestine
Edited by N. C. Gourtsoyiannis

Imaging of the Knee
Techniques and Applications
Edited by A. M. Davies and V. N. Cassar-Pullicino

Perinatal Imaging
From Ultrasound to MR Imaging
Edited by Fred E. Avni

Radiological Imaging of the Neonatal Chest
Edited by V. Donoghue

Diagnostic and Interventional Radiology in Liver Transplantation
Edited by E. Bücheler, V. Nicolas, C. E. Broelsch, X. Rogiers, and G. Krupski

Radiology of Osteoporosis
Edited by S. Grampp

Imaging Pelvic Floor Disorders
Edited by C. I. Bartram and J. O. L. DeLancey
Associate Editors: S. Halligan, F. M. Kelvin, and J. Stoker

Imaging of the Pancreas
Cystic and Rare Tumors
Edited by C. Procacci and A. J. Megibow

High Resolution Sonography of the Peripheral Nervous System
Edited by S. Peer and G. Bodner

Imaging of the Foot and Ankle
Techniques and Applications
Edited by A. M. Davies, R. W. Whitehouse, and J. P. R. Jenkins

Radiology Imaging of the Ureter
Edited by F. Joffre, Ph. Otal, and M. Soulie

Imaging of the Shoulder
Techniques and Applications
Edited by A. M. Davies and J. Hodler

Radiology of the Petrous Bone
Edited by M. Lemmerling and S. S. Kollias

Interventional Radiology in Cancer
Edited by A. Adam, R. F. Dondelinger, and P. R. Mueller

Duplex and Color Doppler Imaging of the Venous System
Edited by G. H. Mostbeck

Multidetector-Row CT of the Thorax
Edited by U. J. Schoepf

Functional Imaging of the Chest
Edited by H.-U. Kauczor

Radiology of the Pharynx and the Esophagus
Edited by O. Ekberg

Radiological Imaging in Hematological Malignancies
Edited by A. Guermazi

Imaging and Intervention in Abdominal Trauma
Edited by R. F. Dondelinger

Multislice CT
2nd Revised Edition
Edited by M. F. Reiser, M. Takahashi, M. Modic, and C. R. Becker

Intracranial Vascular Malformations and Aneurysms
From Diagnostic Work-Up to Endovascular Therapy
Edited by M. Forsting

Radiology and Imaing of the Colon
Edited by A. H. Chapman

Coronary Radiology
Edited by M. Oudkerk

Dynamic Contrast-Enhanced Magnetic Resonance Imaging in Oncology
Edited by A. Jackson, D. L. Buckley, and G. J. M. Parker

Imaging in Treatment Planning for Sinonasal Diseases
Edited by R. Maroldi and P. Nicolai

Clinical Cardiac MRI
With Interactive CD-ROM
Edited by J. Bogaert, S. Dymarkowski, and A. M. Taylor

Focal Liver Lesions
Detection, Characterization, Ablation
Edited by R. Lencioni, D. Cioni, and C. Bartolozzi

Multidetector-Row CT Angiography
Edited by C. Catalano and R. Passariello

MR Imaging in White Matter Diseases of the Brain and Spinal Cord
Edited by M. Filippi, N. De Stefano, V. Dousset, and J. C. McGowan

Paediatric Musculoskeletal Diseases
With an Emphasis on Ultrasound
Edited by D. Wilson

Contrast Media in Ultrasonography
Basic Principles and Clinical Applications
Edited by Emilio Quaia

Springer

Medical Radiology Diagnostic Imaging and Radiation Oncology

Titles in the series already published

Radiation Oncology

Lung Cancer
Edited by C.W. Scarantino

Innovations in Radiation Oncology
Edited by H. R. Withers
and L. J. Peters

Radiation Therapy of Head and Neck Cancer
Edited by G. E. Laramore

Gastrointestinal Cancer – Radiation Therapy
Edited by R.R. Dobelbower, Jr.

Radiation Exposure and Occupational Risks
Edited by E. Scherer, C. Streffer,
and K.-R. Trott

Radiation Therapy of Benign Diseases
A Clinical Guide
S. E. Order and S. S. Donaldson

Interventional Radiation Therapy Techniques – Brachytherapy
Edited by R. Sauer

Radiopathology of Organs and Tissues
Edited by E. Scherer, C. Streffer,
and K.-R. Trott

Concomitant Continuous Infusion Chemotherapy and Radiation
Edited by M. Rotman
and C. J. Rosenthal

Intraoperative Radiotherapy – Clinical Experiences and Results
Edited by F. A. Calvo, M. Santos,
and L.W. Brady

Radiotherapy of Intraocular and Orbital Tumors
Edited by W. E. Alberti and
R. H. Sagerman

Interstitial and Intracavitary Thermoradiotherapy
Edited by M. H. Seegenschmiedt
and R. Sauer

Non-Disseminated Breast Cancer
Controversial Issues in Management
Edited by G. H. Fletcher and
S.H. Levitt

Current Topics in Clinical Radiobiology of Tumors
Edited by H.-P. Beck-Bornholdt

Practical Approaches to Cancer Invasion and Metastases
A Compendium of Radiation Oncologists' Responses to 40 Histories
Edited by A. R. Kagan with the
Assistance of R. J. Steckel

Radiation Therapy in Pediatric Oncology
Edited by J. R. Cassady

Radiation Therapy Physics
Edited by A. R. Smith

Late Sequelae in Oncology
Edited by J. Dunst and R. Sauer

Mediastinal Tumors. Update 1995
Edited by D. E. Wood
and C. R. Thomas, Jr.

Thermoradiotherapy and Thermochemotherapy

Volume 1:
Biology, Physiology, and Physics

Volume 2:
Clinical Applications

Edited by M.H. Seegenschmiedt,
P. Fessenden, and C.C. Vernon

Carcinoma of the Prostate
Innovations in Management
Edited by Z. Petrovich, L. Baert,
and L.W. Brady

Radiation Oncology of Gynecological Cancers
Edited by H.W. Vahrson

Carcinoma of the Bladder
Innovations in Management
Edited by Z. Petrovich, L. Baert,
and L.W. Brady

Blood Perfusion and Microenvironment of Human Tumors
Implications for Clinical Radiooncology
Edited by M. Molls and P. Vaupel

Radiation Therapy of Benign Diseases
A Clinical Guide
2nd Revised Edition
S. E. Order and S. S. Donaldson

Carcinoma of the Kidney and Testis, and Rare Urologic Malignancies
Innovations in Management
Edited by Z. Petrovich, L. Baert,
and L.W. Brady

Progress and Perspectives in the Treatment of Lung Cancer
Edited by P. Van Houtte,
J. Klastersky, and P. Rocmans

Combined Modality Therapy of Central Nervous System Tumors
Edited by Z. Petrovich, L. W. Brady,
M. L. Apuzzo, and M. Bamberg

Age-Related Macular Degeneration
Current Treatment Concepts
Edited by W. A. Alberti, G. Richard,
and R. H. Sagerman

Radiotherapy of Intraocular and Orbital Tumors
2nd Revised Edition
Edited by R. H. Sagerman,
and W. E. Alberti

Modification of Radiation Response
Cytokines, Growth Factors, and Other Biolgical Targets
Edited by C. Nieder, L. Milas,
and K. K. Ang

Radiation Oncology for Cure and Palliation
R. G. Parker, N. A. Janjan,
and M. T. Selch

Clinical Target Volumes in Conformal and Intensity Modulated Radiation Therapy
A Clinical Guide to Cancer Treatment
Edited by V. Grégoire, P. Scalliet,
and K. K. Ang

Advances in Radiation Oncology in Lung Cancer
Edited by Branislav Jeremić

Printing and Binding: Stürtz GmbH, Würzburg

Printing and Binding: Stürtz GmbH, Würzburg